PRACTICAL SHIELDED METAL ARC WELDING

Mike Gellerman

Upper Saddle River, New Jersey Columbus, Ohio

Library of Congress Cataloging-in-Publication Data
Gellerman, Mike
Practical shielded metal arc welding / Mike Gellerman.
p. cm.
Includes index.
ISBN 0-13-601931-5
1. Shielded metal arc welding. I. Title.
TK4660.0385 1998 97-10375
671.5'212—dc21 CIP

Cover photo: Craig Aurness/Westlight
Editor: Ed Francis
Production Editor: Stephen C. Robb
Design Coordinator: Karrie M. Converse
Text Designer: Linda M. Robertson
Cover Designer: Rod Harris
Production Manager: Laura Messerly
Illustrations: Jane Lopez
Marketing Manager: Danny Hoyt

This book was set in Century and Olive by Carlisle Communications, Ltd. and was printed and bound by Banta Company. The cover was printed by Phoenix

Printed in the United States of America

10 9 8 7 6 5 4 3 2 1

ISBN: 0-13-601931-5

Prentice-Hall International (UK) Limited, *London*
Prentice-Hall of Australia Pty. Limited, *Sydney*
Prentice-Hall Canada, Inc., *Toronto*
Prentice-Hall Hispanoamericana, S. A., *Mexico*
Prentice-Hall of India Private Limited, *New Delhi*
Prentice-Hall of Japan, Inc., *Tokyo*
Simon & Schuster Asia Pte. Ltd., *Singapore*
Editora Prentice-Hall do Brasil, Ltda., *Rio de Janeiro*

Preface

The goal or main objective of this book is to develop the ability to do welding safely. This learning material is recommended for use in industrial arts classes, adult education, and vocational and apprenticeship programs by students with a desire to learn basic welding. This book is written to meet some of the specifications set forth in the American Welding Society's *Structural Welding Code—Steel D1.1.* Questions are included throughout. Besides reviewing the content thoroughly, many of the questions have been developed to apply skills such as mathematics and writing, as well as problem-solving skills. Carefully following the teaching provided in this book will enhance your enjoyment of welding and improve your ability to make quality welds.

This book is laid out in an easy-to-understand format for use with an instructor. Gradually, by following the step-by-step approach laid out in the sections of each unit, you will increase your comprehension of the material and master the basics. And the best news is that almost any person who is healthy and has good vision and hand-to-eye coordination can learn basic welding. Students motivated by a conscientious desire to answer the review questions and to complete the welding exercises as often as necessary to meet the criteria established for the class will achieve success in welding.

Welding has its own set of terms. Unfamiliar words will be introduced as the student progresses through this book. Words that appear in **bold type** can be found in the glossary at the end of this book; so try not to skip over words that are unfamiliar—look them up. Finally, make every effort to follow the equipment manufacturers' recommendations for safety. Complete a mental safety checklist each time you set up to do welding. Remember that success at welding will depend on the time and effort you spend to accomplish the given tasks.

Acknowledgments

The author gratefully acknowledges the following reviewers for their insightful suggestions: Andrew M. Burke, Monroe County Community College; Dave Hoffman, Fox Valley Technical College; Carl V. Matricardi, Gwinnett Technical Institute; Ron Whitman, Texas State Technical College.

The author also thanks Dan Maynard, welding instructor at Wisconsin Indianhead Technical College, for his assistance on this project; Industrial Welders and Machinists in Duluth, Minnesota; and Praxair of Superior, Wisconsin. Finally, the author thanks the staff at Prentice Hall, including copy editor Ben Shriver and graphic artist Jane Lopez and especially senior editor Ed Francis, who understood the author's vision for this book.

Brief Contents

Contents

UNIT 1

Welding Theory

1. THE PROCESS OF WELDING

Definition Of Welding

Several processes can join materials together by **welding,** as shown in Figure 1–1. Welding is by definition any process of joining material together by heat. Welding can be likened to carpentry. Both woodworking and metalworking provide the opportunity of putting ideas into practice. Both trades require a plan that usually involves measuring, cutting to size, and fastening together. By learning to weld, many more useful products can be fabricated or repaired using a small reasonably priced welding **power source** (Figure 1–2).

Welding is a method of joining things together that is different from either riveting, bolting, **soldering,** or **brazing,** (see Figure 1–3). In shielded metal arc welding the properties of the **filler metal** are mixed together with those of the **base metal.** Welding is the only method of joining metals in which the filler metal, or fastening material, mixes together so completely with the base metal. The filler metal combines with the base metal to form the **weld.** The weld takes on the characteristics of both the filler metal and the base metal. In most instances the **joint** is stronger than the base metal because the weld itself becomes stronger than the base metal.

Definition of Shielded Metal Arc Welding

Shielded metal arc welding (SMAW) is a welding process that uses an electric **arc** to generate heat. The welding arc is formed between a **flux-covered electrode** and the base metal (Figure 1–4). The electrode, or filler metal, melts into the pool of liquid metal which is called the **weld pool.** The weld pool mixes with

Arc Welding Processes
atomic hydrogen welding AHW
bare metal arc welding BMAW
carbon arc welding CAW
carbon arc welding-gas—CAW-G
carbon arc welding-shielded—CAW-S
carbon arc welding—twin CAW-T
electrogas welding EGW
flux cored arc welding FCAW
gas metal arc welding GMAW
gas metal arc welding-pulsed arc GMAW-P
gas metal arc welding-short circuiting arc GMAW-S
gas tungsten arc welding GTAW
gas tungsten arc welding—pulsed arc GTAW-P
plasma arc welding PAW
shielded metal arc welding SMAW
stud arc welding SW
submerged arc welding SAW
submerged arc welding—series SAW-S

Soldering
dip soldering DS
furnace soldering FS
induction soldering IS
infrared soldering IRS
iron soldering INS
resistance soldering RS
torch soldering TS
wave soldering WS

Solid State Welding
coextrusion welding CEW
cold welding CW
diffusion welding DFW
explosion welding EXW
forge welding FOW
friction welding FRW
hot pressure welding HPW
roll welding ROW
ultrasonic welding USW

Resistance Welding
flash welding FW
projection welding PW
resistance seam welding RSEW
 high frequency RSEW-HF
 inducton RSEW-I
resistance spot welding RSW
upset welding UW
 high frequency UW-HF
 induction UW-I

Brazing
block brazing BB
diffusion brazing DIB
dip brazing DB
exothermic brazing EXB
flow brazing FLB
furnace brazing FB
induction brazing IB
infrared brazing IRB
resistance brazing RB
torch brazing TB
twin carbon arc brazing TCAB

Other Welding
electron beam welding EBW
 high vacuum EBW-HV
 medium vacuum EBW-MV
 nonvacuum EBW-NV
electroslag welding ESW
flow welding FLOW
induction welding IW
laser beam welding LBW
percussion welding PEW
thermit welding TW

Oxyfuel Gas Welding
Air acetylene welding AAW
oxyacetylene welding OAW
oxyhydrogen welding OHW
pressure gas welding PGW

FIGURE 1–1 Chart of welding processes. Shielded metal arc welding is only one of eighteen arc welding processes.

the base metal, the material being welded. As the weld cools, it is protected by gases given off by the flux covering. These gases stabilize the arc, as well as keeping oxygen and nitrogen in the air away from the weld pool. The flux covering, as it breaks down, picks up impurities and forms a protection of **slag** over the weld. The completed weld, acting like a fastener, joins metal together in a rigid structure. Using this process, the completed weld is the result of melting metal together in a joint with filler metal.

Description Of Shielded Metal Arc Welding

Shielded metal arc welding is a relatively inexpensive welding process requiring a minimum amount of equipment. In this welding process an electric arc is established between the tip of a covered electrode and the base metal. The intense heat of the arc

FIGURE 1–2 Transformers are very popular for the home workshop. Lincoln AC/DC and AC power sources.

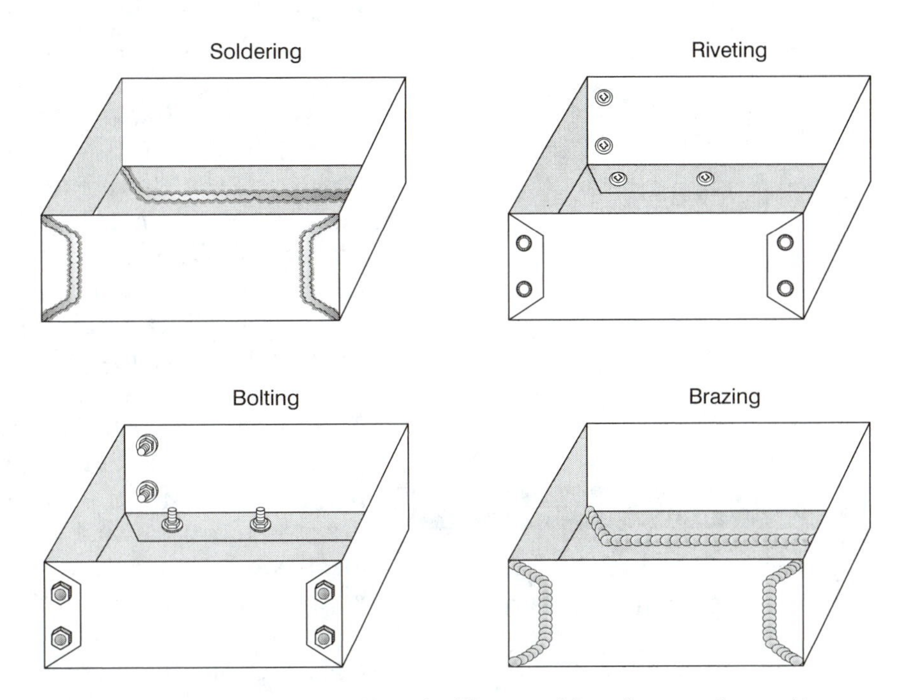

FIGURE 1–3 Soldering, riveting, bolting, and brazing are four other methods of joining metals together.

melts the electrode, which fuses with the metal of the joint. Temperatures in the arc have been measured in excess of 9000° F.

The arc is formed at the gap in the electric circuit. The gap is made by touching the tip of the electrode to the base metal, then raising the electrode slightly. This causes electric current to arc by

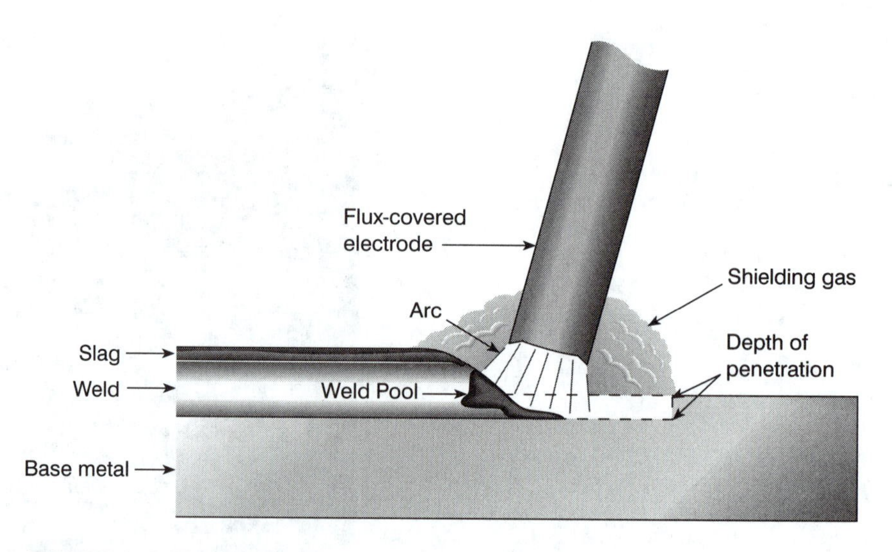

FIGURE 1–4 The welding arc. The weld pool is protected from the air by a shielding gas produced by the flux covering, which becomes slag and continues to protect the weld as it cools.

jumping the gap. **Input** current is delivered to the power source by the electric power company. The input current is converted to **output** current for the purpose of welding. The output current moves through the **workpiece connection** (ground clamp) and the **electrode holder** (Figure 1–5).

Shielded metal arc welding requires a **constant current** power source. This means that from an open circuit of between 50 and 100 volts there is a steep drop to 25 volts during welding (see Figure 1–6). This higher open-circuit voltage is one reason why it can be so easy to put a hole through thin metal upon starting an arc. There is just a brief moment of time in which to establish a lower welding voltage.

The **quality** of the weld is affected by the length of the arc. The welder has control over the length of the arc. The welder must learn how to maintain the proper **arc length** while moving along the joint in the act of welding, as shown in Figure 1–7. Too long an arc length can cause a problem in which molten metal from the electrode is thrown away from the weld. This problem is called **spatter.** Too short an arc length results in sticking the electrode to the base metal, putting out the arc (Figure 1–8). Unit 4 will cover the subject of arc length in greater detail.

ARC FLASH

After completing a project, the welds should be painted to protect them from rusting.

SAFETY REMINDER

The shop should be equipped with one or more fire extinguishers placed in designated locations. Know the classification of each fire extinguisher for correct application. See Unit 2 for the four types of fire extinguishers.

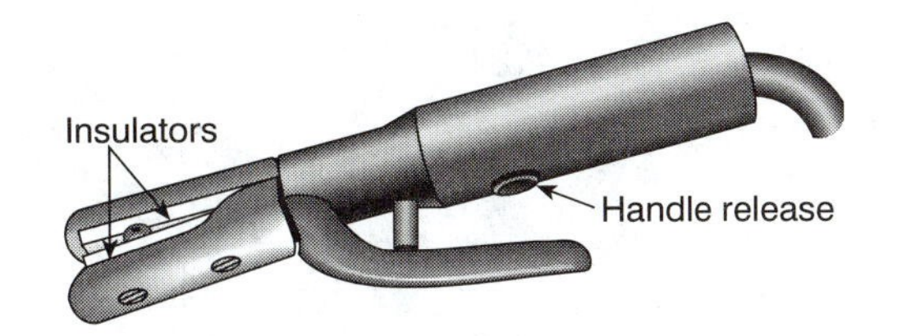

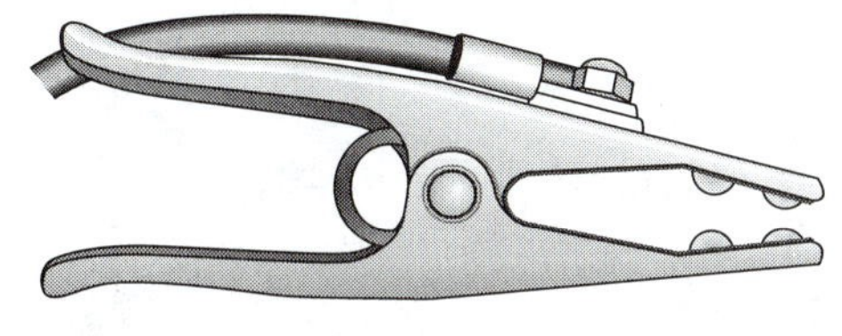

FIGURE 1–5 Workpiece connection and electrode holder. The insulators keep the holder from arcing off the base metal.

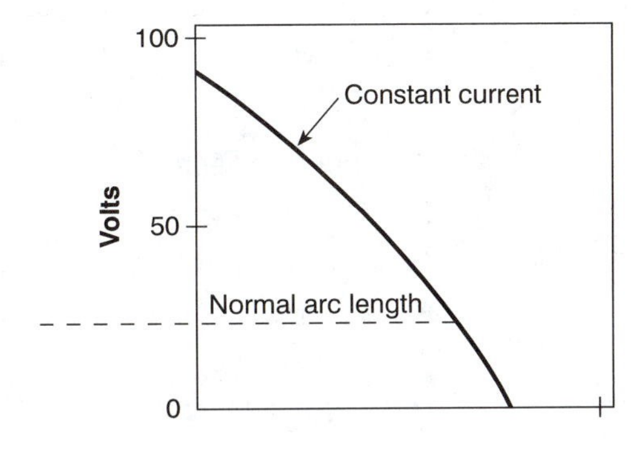

FIGURE 1–6 Typical voltage drop during welding. Normal arc voltage can range from 17 to 27 V depending on size of the electrode.

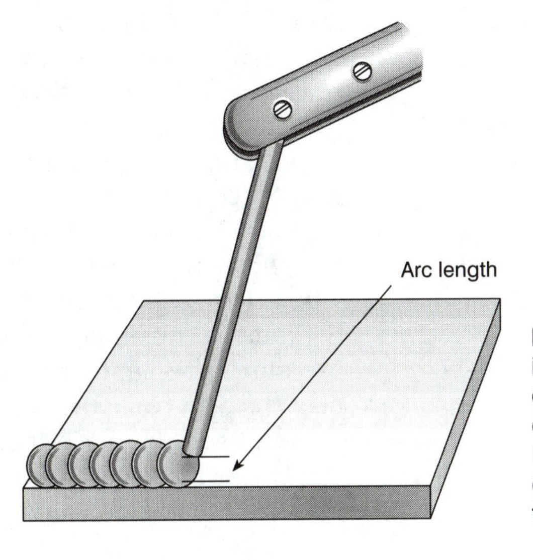

FIGURE 1–7 Arc length is important in welding, giving the welder some control over arc voltage by the distance the electrode is held above the base metal.

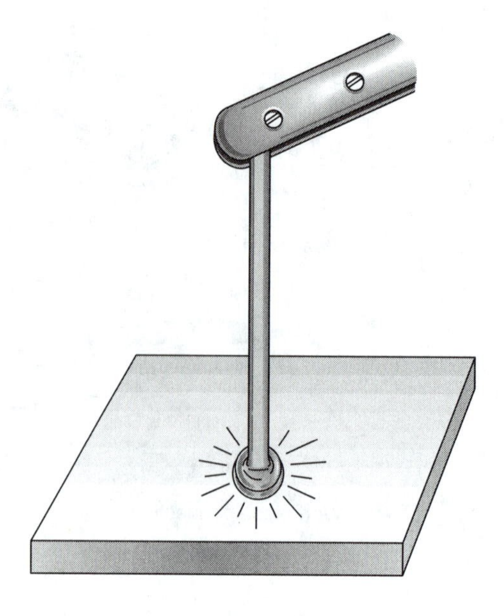

FIGURE 1–8 Improper arc length is one reason for sticking the electrode to the base metal.

REVIEW

1. How is welding related to carpentry?
2. What happens to the electrode during the welding process?
3. What is the purpose of a weld?
4. Describe what happens in the arc.
5. What problems are caused by an improper arc length?

2. ELECTRIC WELDING

Basic Electric Theory

Shielded metal arc welding begins by turning on the electric power. **Current, voltage,** and **amperage** are three terms at the foundation of electrical theory. Current is flow. Amperage is measurement of current flow. Voltage is measurement of the force or pressure causing the flow of electricity.

Using a garden hose to illustrate, current is like the amount of water, and voltage is like the pressure of the water in the hose. Amperage is like the amount of water flowing through the hose measured in gallons. If the amperage setting is too low, like a dripping hose with low water pressure, not enough current is provided. The result can be a weld with incomplete penetration into the joint. If the amperage setting is too high, like a fire hose with pressure too high to water a garden, excessive current can overheat the electrode, breaking down its flux covering while causing numerous problems in the arc.

The welder must control the voltage and the amperage to make a quality weld. This voltage and amperage become the output current in the arc during welding. This output current can vary depending on the arc length. Remember that each weld bead is affected by how well the arc is kept under control in the act of welding.

Input Current

Alternating current (AC) and **direct current (DC)** are used in electric welding. *Alternating* is the term describing the back and forth motion of the current. The current changes directions 120 times per second within an electric circuit. Each back and forth motion of current is called a cycle. The standard **frequency** in the United States is 60 cycles per second. AC supplies the input current to power familiar household appliances such as the electric shaver and the hair dryer. Important in welding, AC can be the output current used to generate thermal energy or heat for shielded metal arc welding (see Figure 1–9).

DC, or direct current, is the second variation of electric power. Unlike AC, or alternating current, DC is current flow in one direction. DC is used to operate electrical systems on automobiles, motor homes, and power boats, to name just a few applications. Most importantly, DC can be the output current used to generate heat for shielded metal arc welding by using a **rectifier** (Figure 1–10). A rectifier converts AC to DC.

The power source used in welding converts high-voltage and low-amperage AC input current into low-voltage and high-amperage output current necessary for welding. This output current can be converted to DC by use of a rectifier. Remember that the input current going into the power source for welding should be serviced only by a qualified electrician.

FIGURE 1–9 The transformer power source is available in different styles. A crank on top is used to adjust the amperage setting. *(Courtesy of Miller Electric Manufacturing Company, Appleton, Wisconsin.)*

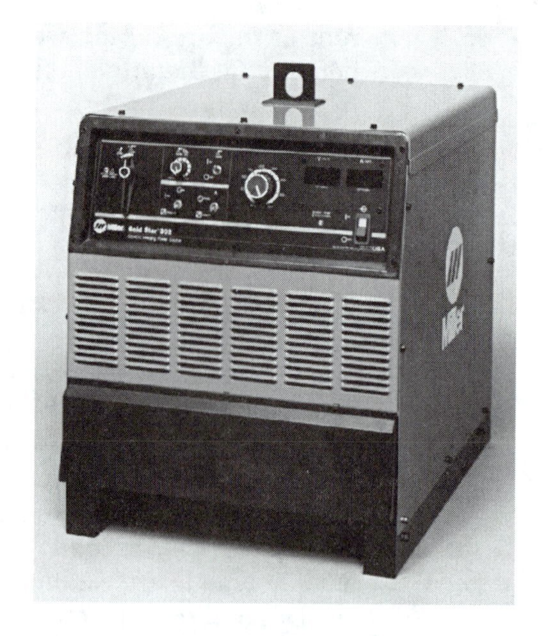

FIGURE 1–10 DC transformer-rectifier power source. *(Courtesy of Miller Electric Manufacturing Company, Appleton, Wisconsin).*

SAFETY REMINDER

Be sure the shop is clear of **volatile** and **flammable** materials before making any attempt at welding.

REVIEW

1. Use the following words in a paragraph: *current, voltage, amperage, flow, pressure,* and *measurement.*
2. How does input current differ from output current?
3. What is AC, or alternating current?
4. What is DC, or direct current?
5. What is the purpose of a rectifier?

3. OUTPUT CURRENT

Choosing a Current

Three choices of output current are available in shielded metal arc welding (and welding is affected by the choice of current). AC is one choice. Because the current is constantly reversing directions during AC welding, the arc goes out and restarts itself with every change of directions, as pictured in Figure 1–11. The arc goes out 120 times a second. This means with AC the arc is not as stable and is slightly more difficult to start. The constant change of direction by AC can reduce the strength of the magnetic field created by the flow of current. AC reduces **arc blow,** a problem that commonly occurs during DC welding. Arc blow is an erratic arc

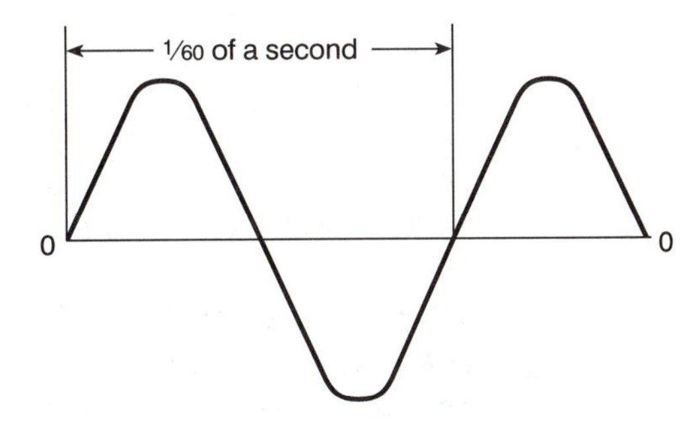

FIGURE 1–11 The wave cycle shows zero where the arc goes out and restarts with every change of direction. Changing direction 120 times per second, AC choice of welding current is not as stable as DC.

FIGURE 1–12 Power source set up for DCEP (direct current electrode positive). The electrode holder is attached to the positive terminal. For DCEN (direct current electrode negative), the electrode holder is attached to the negative terminal.

that refuses to go where the welder desires. See Unit 5 for details concerning arc blow.

The other two choices of current are **DCEP (direct current electrode positive)** and **DCEN (direct current electrode negative).** With DCEP welding, the electrode holder is attached by the electrode **lead** to the positive terminal, and the workpiece connection is attached to the negative terminal on the power source (see Figure 1–12). With DCEN, the electrode holder is attached by the electrode lead to the negative terminal, and the workpiece connection is attached to the positive terminal on the power source. Power sources can be converted from DCEP to DCEN by manually exchanging terminal connections or by flipping a switch.

During DC welding the arc is usually more stable unless arc blow becomes a problem. DCEP results in deeper **root penetration** than either AC or DCEN. Root penetration is the depth of penetration into the joint. With DCEP, maximum heat is released at the base metal side of the arc to cause deeper root penetration. The case is just the opposite during welding with DCEN in which the melting rate increases on the electrode side of the arc due to maximum heat release, causing minimum root penetration. DCEN is used for the welding of thin metal that may require minimum root penetration. AC has a root penetration between DCEP and DCEN. In AC welding, 50 percent of the heat is released at the electrode side of the arc, and 50 percent of the heat is released at the base metal side of the arc.

Understanding the differences among the three choices of current available for welding is important. The welder should learn to match the current with the conditions of the welding job.

REVIEW

1. Explain the relationship between the following sets of terms:
  a. *AC* and *arc blow*
  b. *DCEP* and *electrode holder*
  c. *DCEN* and *workpiece connection*
  d. *DCEP* and *root penetration*
  e. *DCEN* and *root penetration*

Lab Exercise

1. Examine a power source in the shop, and complete the following:
  a. What choices of current does it provide?
  b. How does the power source convert from one current to another?
  c. If possible, switch the power source from AC to DC.
  d. If possible, switch the power source from DCEP to DCEN.

REVIEW QUESTIONS FOR UNIT 1

Multiple Choice

Choose the best answer to complete the statement.

1. Welding is like:
  a. Riveting and bolting.
  b. Bolting and soldering.
  c. Brazing and soldering.
  d. Brazing and riveting.
  e. All of the above.

2. Filler metal:
  a. Forms a flux covering.
  b. Melts into the weld pool.
  c. Is nonmagnetic.
  d. Becomes slag.
  e. None of the above.

3. Flux:
  a. Produces gases that protect the weld.
  b. Forms a slag.
  c. Stabilizes the arc.

d. Breaks down impurities.
e. All of the above.

4. The arc:
a. Joins the electrode with the base metal.
b. Is not part of the electrical circuit.
c. Melts the electrode.
d. Is created by turning on the power source.
e. None of the above.

5. Input current:
a. Is converted to output voltage.
b. Is around 25 volts.
c. Moves through the circuit consisting of the workpiece connection and the electrode holder.
d. Is delivered to the power source.
e. None of the above.

6. Alternating current:
a. Produces an arc that is very easy to start.
b. Flows in one direction.
c. Is simply the back and forth motion of current.
d. Operates the electrical system on automobiles.
e. Requires a rectifier.

7. With DCEP,
a. The electrode is attached to the workpiece connection.
b. Arc blow is less of a problem than with AC.
c. Maximum heat is released on the electrode side of the arc.
d. You are using a current that is preferred on thin metal.
e. None of the above.

8. ______________ is the length between the electrode and the base metal.
a. Open-circuit voltage
b. Constant current
c. Voltage
d. Arc length
e. None of the above

9. Welding:
a. Is a trade.
b. Consists of many different processes.
c. Is any process of joining materials by heat.
d. Is different from brazing.
e. All of the above.

Short Answer

1. What is shielded metal arc welding?
2. How is input current different from output current?
3. What is the arc?
4. What three terms provide the foundation for the theory of electricity?
5. What is one result from too long an arc length?
6. What is one result from too short an arc length?
7. How is alternating current different from direct current?
8. What current is more likely to result in arc blow?

9. What does DCEP mean?
10. What are some of the differences between AC and DC welding?

ADDITIONAL ACTIVITIES

Riddles

Solve the following teasers.

1. I am a measurement of force causing the flow of electricity.
2. I am the molten metal from the electrode thrown away from the weld pool.
3. I'm the junction of two or more pieces joined together by welding, brazing, or soldering.
4. I am a measurement of the rate of current flow.
5. I am nothing more than the weld from a single pass.
6. I am the material that is welded.
7. Without me the base metal could not be connected to the power source.
8. Watch out for me, for I am capable of causing an explosion.

Writing Skills

1. Write out the shop procedures to be followed in case of an accident.
2. Write a paragraph using all the following terms in sentences: *weld bead; arc length, spatter, amperage, DCEP.*
3. Write a paragraph expressing your thoughts on welding at this point in the program.

Math—Decimals and Percentages

Understanding the composition of steel—its carbon content, for example—requires a knowledge of decimals and percentages.

0.1 = one tenth	0.01 = one hundredth	0.001 = one thousandth
1% = one hundredth	10% = one tenth	100% = one

Remember that to change a percentage to a decimal, drop the percent sign (%) and add two decimal places. Example: 1% = 0.01

Solve for the decimals.

1. 40% =	**2.** 4% =	**3.** 0.4% =	**4.** 7% =	**5.** 80% =
6. 0.7% =	**7.** 70% =	**8.** 7% =	**9.** 3.5% =	**10.** 4.5 %=

To convert a decimal to a percentage, move the point two decimal places to the right and add the percent sign. Example: 0.1 = 10%

Solve for the percentage.

11. 0.03 =	**12.** 0.3 =	**13.** 0.35 =	**14.** 0.04 =	**15.** 0.045 =
16. 0.7 =	**17.** 0.07 =	**18.** 0.007 =	**19.** 0.25 =	**20.** 0.025 =

Solve for the expression in words. Example: 0.1 = one tenth

21. 0.4 = **22.** 0.045 = **23.** 0.07 = **24.**0.075 = **25.** 0.02 =

Puzzlers

1. The power source is on and running, and the workpiece connection and the electrode holder are both in good condition; in fact, the welder was welding on another project just five minutes ago. However, now the welder is unable to strike an arc. The problem is not in a bad circuit from the workpiece connection or the electrode holder to the power source. What are three other possibilities?
2. Your neighbor just ran his car over a new piece of lawn furniture and would like to see if the neighborhood welder can do the repair. He thinks the job shouldn't take a minute of the welder's time. The chair, still smelling of a fresh white paint job, has a crack in a ⅛-inch thick steel bracket attached to the leg. What problems could the welder have by immediately attaching the workpiece connection and beginning to weld up the crack?

UNIT 2

Welding Equipment

1. SAFETY IN WELDING

Safety First

Safe electric welding practices will prevent injuries. Develop an attitude about safety in everthing you do. Very serious injury can happen to even experienced welders when they become careless. Think safety.

Shielded metal arc welding produces sparks hot enough to start a fire. A welder should know the type of fire extinguisher required to put out a given fire. There are four main types of fire extinguishers.

Type A fire extinguishers are for use on combustible materials such as cardboard, wood, and paper (see Figure 2–1). Type A is recognized by the symbol of a green triangle with the letter A in the center.

The type B fire extinguisher is for use on **volatile** substances including gas, oils, paints, and solvents, as shown in Figure 2–2. Type B is denoted by the symbol of a red square with the letter B in the center.

A type C fire extinguisher is for use on electrical fires such as those in fuse boxes, motors, and wiring (Figure 2–3). Type C is marked with the symbol of a blue circle with the letter C in the center.

Type D fire extinguishers are for use on metal fires that include magnesium, titanium, and zinc, as in Figure 2–4. Type D is recognized by the symbol of a yellow star with the letter D in the center.

Smoke, light, heat, sparks, and shock are five areas of concerns with shielded metal arc welding (Figure 2–5). The smoke given off by welding can contain fumes and gases that may affect breathing if inhaled over a period of time. **Ultraviolet** light given

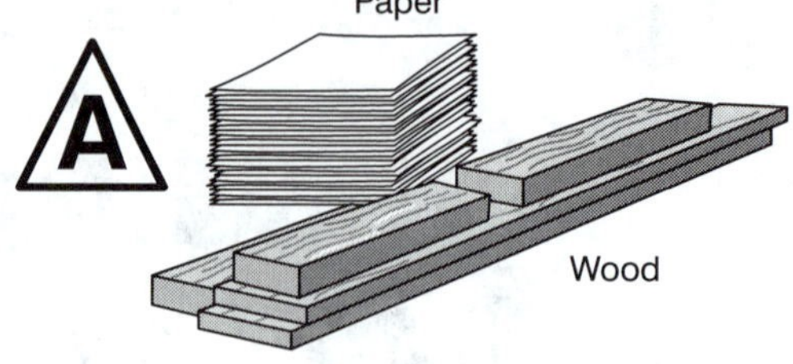

FIGURE 2–1 Type A fire extinguisher. The symbol of a green triangle is for combustible materials such as paper and wood.

FIGURE 2–2 Type B fire extinguisher. The symbol of a red square is for volatile substances such as gas, oil, and paint.

FIGURE 2–3 Type C fire extinguisher. The symbol of a blue circle is for electrical fires.

FIGURE 2–4 Type D fire extinguisher. The symbol of a yellow star is for metal fires such as burning magnesium, zinc, and titanium.

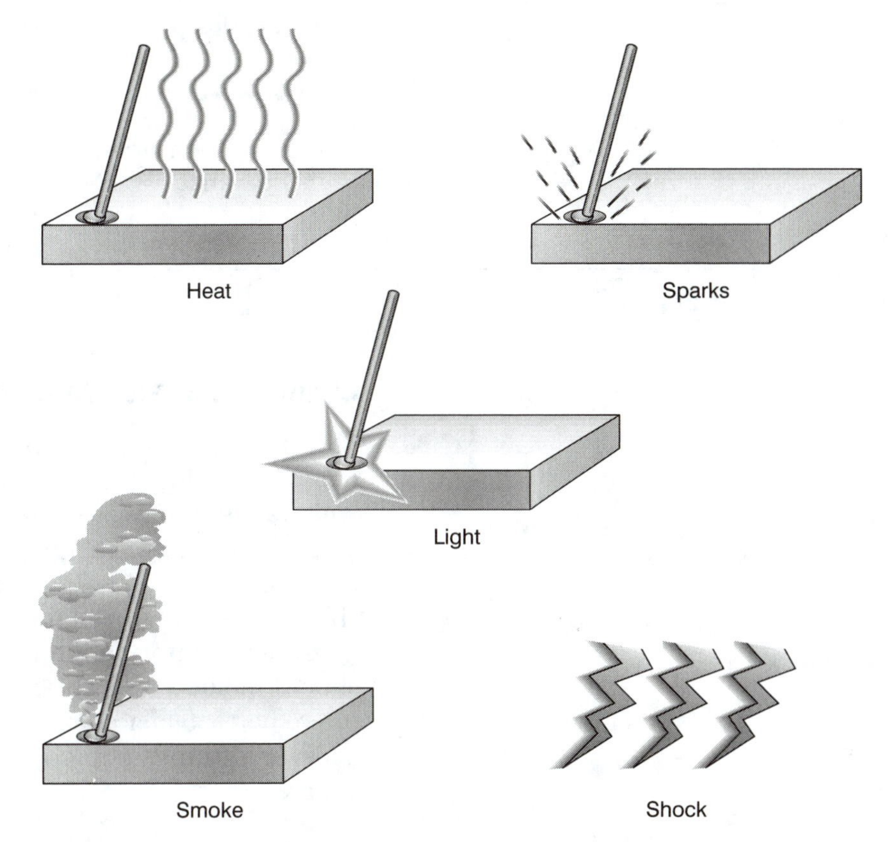

FIGURE 2–5 The five major dangers associated with shielded metal arc welding: heat, sparks, light, smoke, and shock.

off by the welding arc can affect the eyes. Heat exposure to the skin produced by welding can cause from first-degree to the most severe third-degree burns. Sparks given off by welding will destroy clothing, damage the skin, start **flammable** materials on fire, and cause volatile substances to explode. Electric shock can cause injury and even death.

In describing the equipment for shielded metal arc welding, attention will center on safety in these five areas of concerns. Using the proper welding equipment and supplies is essential for your own protection. Pay attention to the suggested safety guidelines listed at the end of this unit. A welder should constantly be aware of the dangers presented by smoke, ultraviolet light, heat, sparks, and shock. Always take the necessary safety precautions, and remember your ultimate safety depends on you.

REVIEW

1. What should be known about the selection of a fire extinguisher?
2. Explain the relationship between the following sets of terms:
   a. smoke and breathing
   b. light and the eyes
   c. heat and skin
   d. sparks and fire
   e. shock and injury
3. What is the most important idea about safety?

Lab Exercise

1. Identify each type of fire extinguisher in the shop and its recommended use.

2. POWER SOURCES

Choice of Welding Machines

The **power source** is the welding machine used for shielded metal arc welding. Developments in technology have produced a wide range of power sources from which to choose. Selection of a power source should depend on the requirements of the particular welding process. This section will focus on the process of shielded metal arc welding.

A list of information to consider in choosing a power source should include repair requirements, portability, safety, purchase price, parts availability, and electrical code requirements. Three additional things should be considered in deciding on a power source:

1. Does the welding power source operate with single or three-phase input power? You want to match the requirements of the welding power source with the input power available through the electric company.
2. Is the welding power source of sufficient amperage (amp) for the job requirements?
3. What type(s) of current output does the power source provide: AC, DC, or both?

Schools and industrial welding applications utilize input current provided by three-phase power. Three-phase power is less expensive to operate, though more expensive to install. On the other hand, household current for the home workshop generally runs on single-phase input power. If a power source is to operate on single-phase power, then a model with a voltage input of 230 volts

FIGURE 2–6 Single-phase power source with input power of 115 volts. A welding machine that is available for light welding applications. (Courtesy of ESAB Welding and Cutting Products, Florence, South Carolina.)

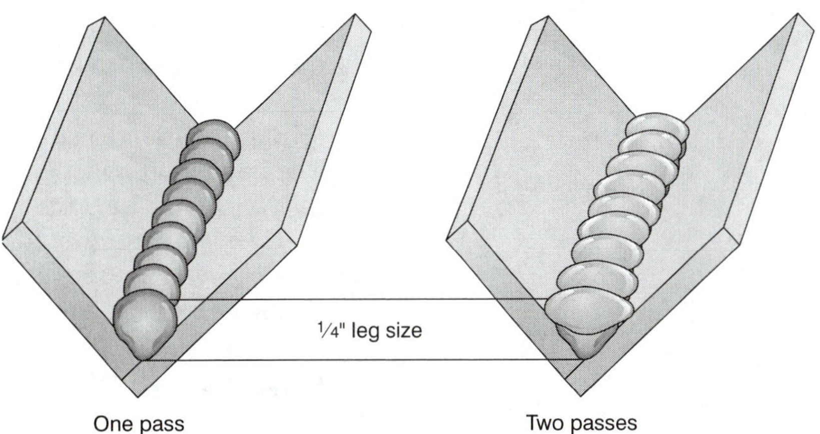

FIGURE 2–7 One pass can result in less time but more distortion.

is readily available (Figure 2–6). This is the same input voltage necessary to operate the kitchen stove.

The selection of a power source based on the job requirements should consider the thickness of the base metal. A $\frac{1}{4}$-inch fillet weld will join two pieces of $\frac{3}{8}$-inch thick plate in the flat position with one pass, using a $\frac{1}{4}$-inch diameter electrode at a setting of approximately 350 amperes. The same job can be completed in two passes using the $\frac{1}{8}$-inch diameter electrode at a setting of 130 amperes (see Figure 2–7). Welding takes more time with a smaller power source requiring smaller electrodes, although less heating of the joint results in less distortion.

The minimum output amperage for schools and industrial welding applications will vary depending on particular applications. Output amperage for welding applications in most home workshops should be a power source with a minimum 180 amperes AC/110 amperes DC. A qualified electrician should make any of the wiring changes necessary to accommodate the installation of any power source for welding. The home or shop insurance policy should be examined to be sure of coverage.

The minimum electrical information for rating power sources as determined by the National Electrical Manufacturers Association

1. Manufacturer's name	
2. Class 1 60%, 80%, or 100% duty cycle Class II 30%, 40%, or 50% duty cycle Class III 20% duty cycle	
3. Input voltage	**4.** Input amperage
5. Maximum open circuit voltage (OCV)	
6. Rated load voltage	
7. Rated load amperage	
8. Phase	**9.** RPM at no load
10. Duty cycle	
11. Frequency	

FIGURE 2–8 Information found on the nameplate is useful in giving the capabilities of a power source.

(NEMA) is stated on the nameplate of the machine (see Figure 2–8). The nameplate must be attached to the power source. This information should be brought to the attention of the electrician. Under no circumstance should a welder ever remove the outer covering of the power source. Any servicing of the internal components of the power source must be left to a qualified electrician.

Duty Cycle

Electrical components heat up under the load of welding. And unless the power source is designed for continual welding, it is intended to deliver power for a limited time period. So each power source is rated with a **duty cycle.** A duty cycle is the recommended percentage of time a power source should be under the load of welding. A 30% duty cycle means the power source is rated for welding 3 minutes out of every 10 at its rated output. Rated output is usually, but not always, the maximum amperage setting. For welding loads under the maximum, the duty cycle should not be a problem. Consider this example of finding the maximum amperage setting in a given situation.

> The welding machine is a 300-amp power source with a 60% duty cycle. A 60% duty cycle means the power source is rated for welding 6 minutes out of every 10 at the maximum amperage setting. Maximum amperage in this case is found by multiplying 300 by $\sqrt{.60}$, which equals 232. An answer of 232 means continuous welding at 232 amps or below is safe.

While the chance of the power source overheating is unlikely, sometimes it is necessary to set up a power source with a 60% duty cycle to weld continually (100%). Sometimes too, a power source is required to produce more than its rated output amperage. In these situations being able to do calculations based on the duty cycle is important.

Five Types of Power Sources

The **transformer** is an inexpensive AC power source. Using it, high-voltage and low-amperage input power is transformed into the low-voltage and high-amperage output power necessary for welding. The transformer power source is normally single-phase. The National Electrical Manufacturers Association (NEMA) Class III transformer power source rated with a 20% duty cycle is popular on farms and in shops where some repair and fabrication welding is required.

A **rectifier** added to a transformer changes AC to DC (Figure 2–9). The rectifier is an additional cost to the purchase price of a power source. However, the added DC capability provides greater versatility since some electrodes will not operate in AC. While this type of power source supplies both AC and DC welding capabilities, some rectifier power sources are designed strictly for DC welding.

The **inverter** is a third type of power source. The latest development in technology, these lightweight welding power sources can weigh less than 50 pounds. Inverters are also manufactured with either single-phase or three-phase input power in one unit. Some inverters have been designed with the capability of performing several other welding processes in addition to shielded metal arc welding (see Figure 2–10).

The gasoline or diesel engine driven is a fourth type of power source. This power source can be easily transported with very little maintenance for operation in extreme weather without fear of breakdown or concern for the welding environment. Where electricity is not available, an engine driven power source can be dropped into the most remote location to perform the necessary welding. Logging equipment that breaks in a difficult to reach forest area can be repaired on the spot by an engine driven power source (Figure 2–11).

The motor generator is the fifth type of power source. The motor generator is similar to the engine driven power source, but an electric motor rotates an armature to produce the current for welding (Figure 2–12). With a motor generally wired as three-phase, a

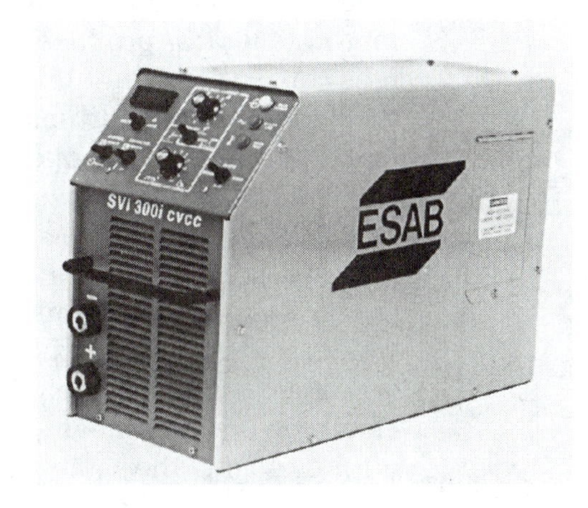

FIGURE 2–9 A versatile power source designed not only for shielded metal arc welding, but gas metal arc welding, gas tungsten arc welding, and air carbon arc gouging. (Courtesy of ESAB Welding and Cutting Products, Florence, South Carolina.)

FIGURE 2–10 Inverter power source. The XMT 304 is multiprocess. This power source has the capability for shielded metal arc welding, gas metal arc welding, and gas tungsten arc welding. (Courtesy of Miller Electric Manufacturing Company, Appleton, Wisconsin.)

FIGURE 2–11 An engine driven power source can be hauled by trailer or set on the back of a truck. (Courtesy of Vista Equipment Company, Inc., Crawfordsville, Indiana.)

variable-voltage motor generator gives the welder more control over the welding parameters. The current can be set for a given voltage and the voltage adjusted as shown in Figure 2–13 for large electrodes, vertical position, and overhead position welding. The motor generator produces very smooth arc characteristics and is noisy but so very durable that some have given steady service since the early 1940s. Unfortunately, besides being noisy the motor generator is expensive to operate.

The choice of a power source for shielded metal arc welding can sometimes be reduced to its cost. Cost can vary from a few hundred dollars to several thousand dollars for an engine driven power source. A few hundred dollar power source can do the welding of the more expensive units, but will require more time welding. Less expensive power sources demand smaller diameter electrodes, which means that less filler metal is being deposited during roughly the same amount of welding time.

FIGURE 2–12 Motor generator power source.

FIGURE 2–13 Control panel for motor generator power source.

A welding power source should provide years of service without the worry of repair problems; that is, it should be durable provided the owner is careful to follow the manufacturer's operating instructions. Some transformer power sources have been operating on a regular basis for over forty years of trouble-free service.

SAFETY REMINDER

Any piece of electrical equipment used in a shop can cause injury—a grinder, a drill, a sander, or a power source used for welding. Thoughts about safety should be foremost in your mind at all times. Also, do not operate shop equipment without wearing both **safety glasses** and **ear plugs.**

LAB EXERCISES

1. Examine a power source for shielded metal arc welding and find out the following information:
   a. Manufacturer and brand name
   b. Type of power source
   c. Input phase power
   d. Output current choices
   e. Amperage output range
2. Copy down the electrical information given on the nameplate.
3. Calculate (from the duty cycle of the power source) the amperage setting for continuous welding.
4. Does the shop have a gasoline, diesel, or motor generator? If so, explain how you know.
5. If the shop has both a motor-generator and a transformer/rectifier power source, examine their control panels.
   a. How are they different?
   b. How are they similar?

3. SAFETY EQUIPMENT

Safety Glasses

No shop work should be performed without the use of **safety glasses** (Figure 2–14). Safety glasses are not sufficient unless they contain side shields. Besides offering protection from normal shop activities, safety glasses protect the eyes from slag (a non-metal by-product of the weld pool). Protect the most precious of your five senses, and never wear safety glasses without a welding helmet and the required filter lens to protect against **arc flash,** the extremely bright light produced in welding.

ARC FLASH

A magnifier can be purchased that fits into the helmet to aid those with eyesight problems, including wearers of bifocals.

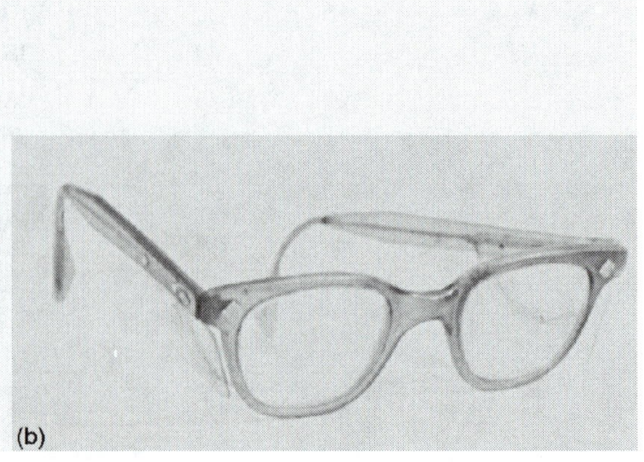

FIGURE 2–14 Safety glasses with side shields. Safety glasses come in many styles, but all of them must protect the eyes from hazards threatening from the front and the sides.

Ear Plugs

Ear plugs are probably the most overlooked piece of safety equipment (see Figure 2–15). Ear plugs may be neglected because hearing loss develops so gradually. Wearing the appropriate ear plugs helps protect the ears from damaging high-frequency sound waves produced by grinding. Under the decibel scale for measuring sound, the threshold of pain is 140 decibels. Sustained exposure from between 120 to 140 decibels can cause hearing loss.

Ear plugs also prevent hot slag and sparks from entering the ear. Do not take this small piece of safety equipment for granted. Use them and continue to enjoy all the wonderful sounds of our world.

Respirators

The respirator is required whenever there is a need to reduce exposure to airborne contaminants you might breath in that are potentially harmful to the body. The base metal, its different types of coatings, electrodes, and objects in the welding area all can produce fumes and gases in the smoke given off by welding. Filter and cartridge respirators are one type of air-purifying respirator. The air-purifying respirator stops contaminants in the air from passing through the nose and the mouth. They are available in hardware stores (Figure 2–16).

The atmosphere-supplying respirator is the second type respirator. Whatever the type, an air-purifying respirator is necessary when the fumes and gases produced by welding cannot be breathed or the welding is being performed in a confined area. If

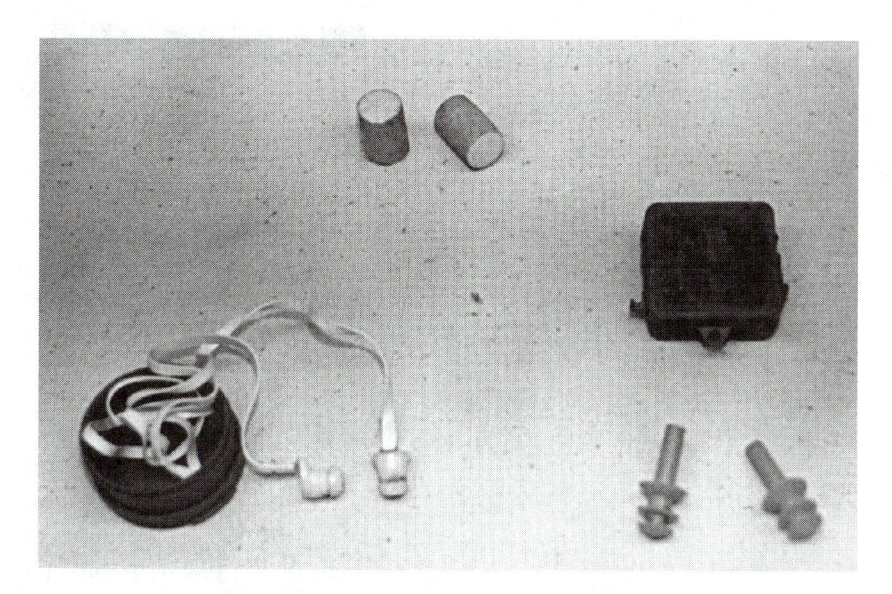

FIGURE 2–15 Ear plugs protect hearing. Choose a pair from the various designs available that are comfortable to wear, and wear them.

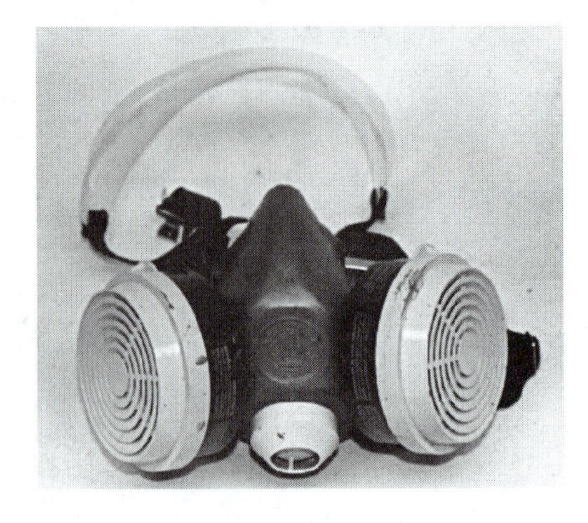

FIGURE 2–16 An air-purifying respirator must be worn in certain environments where there is a danger of inhaling toxic materials.

FIGURE 2–17 Welding helmet with flip-up lens. This type allows the welder to examine the work without having to flip down the entire helmet.

SAFETY REMINDER

Safe practice requires that welders always try to keep their heads out of the smoke stream during welding.

there is ever a question as to the danger from breathing the smoke produced, do not perform the welding without the required protection.

Welding Helmet

The welding helmet (Figure 2–17) protects the head, the face, and the eyes from the effects of the arc during welding. There are many styles of helmets available on the market. A lightweight helmet is best to avoid neck strain. A helmet with a flip-up filter lens (2-inch by 4½-inch) offers better accuracy in placement of the electrode before critical welding. However, a helmet designed

FIGURE 2–18 Welding helmet in which light windows darkens instantly upon arc strike. This type of helmet is expensive.

Table 2–1 Shielded metal arc welding filter lens chart for lenses made from DESAG welding protection glass.

Protection against Ultraviolet and Infrared Radiation (exceeds ANSI Z87.1)	
Shade Number	*Amperage Setting*
9	20 to 39 amps
10	40 to 79 amps
11	80 to 174 amps
12	175 to 299 amps
13	300 to 499 amps

with a larger sized filter lens (5¼-inch by 4½-inch) that does not flip up increases the field of vision. Some expensive helmets are available that convert the lens to the necessary filter instantly upon making an **arc strike.** One is shown in Figure 2–18.

Make it a habit to hold the helmet and the filter lens up to the lights for inspection before welding. A crack in the helmet or the filter lens should be given immediate attention, it should be replaced. The filter lens should meet the shade number for the amperage setting to be used during welding. See Table 2–1 for suggestions on the shade number of the filter lens most appropriate

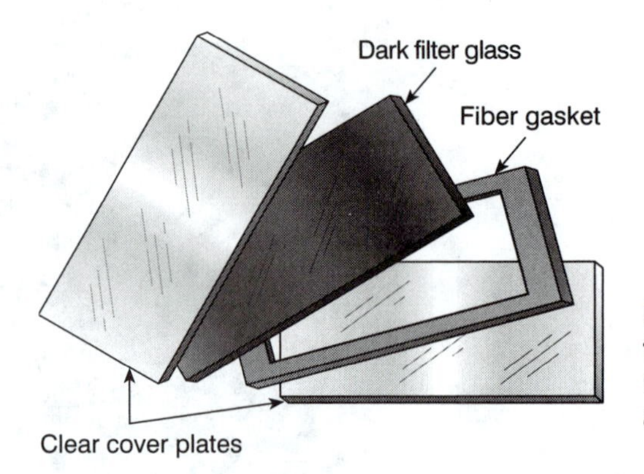

FIGURE 2–19 The entire filter lens assembly should be disassembled and kept clean.

SAFETY REMINDER

As a common courtesy, alert anyone in the immediate area before making an arc strike so eyes can be protected from the flash of the arc.

for shielded metal arc welding. The filter lens is designed to absorb the infrared and ultraviolet light produced by welding. Follow the lens manufacturer's recommendations to match the filter lens with the welding process. Use as dark a lens as possible—and no lighter—to observe both the weld pool and the joint at the point of welding.

Be sure to examine the entire filter lens assembly before welding. While checking for cracks, clean the assembly of smoke stains. For some style helmets the filter lens is sandwiched between two clear cover plates. The cover plate nearest the eyes should be shatter resistant. A fiber gasket separates the cover plate on the arc side of the filter lens, as shown in Figure 2–19. The filter lens should always be installed according to the manufacturer's recommendations.

Arc flash is a painful and usually temporary eye condition caused by the light of the welding arc. Never weld without a helmet, and protect your eyes from the welding arcs of others who may be working in the vicinity. A flash curtain or some type of barrier should be used to screen onlookers.

Welding Gloves

Welding gloves protect the hands and lower arms from burns and electric shock. Only gloves made for welding do an adequate job of handling heat without losing their shape. Gloves with gauntlets that protect the wrists and forearms should be worn (Figure 2–20). Use tongs to handle hot metal; the welding gloves will last longer and the leather will not become stiff; see Figure 2–21.

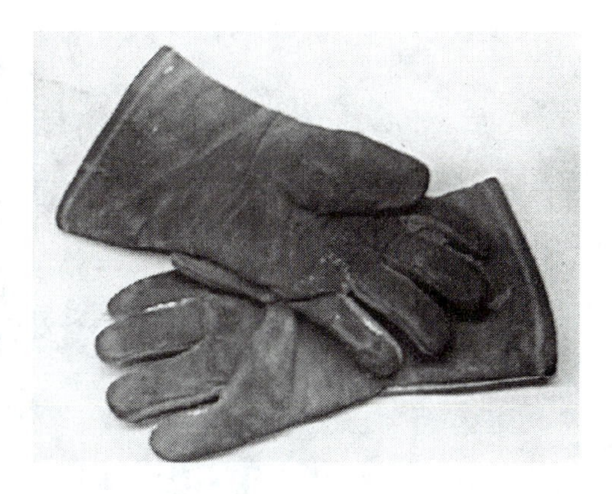

FIGURE 2–20 Welding gloves with gauntlet protection should be worn.

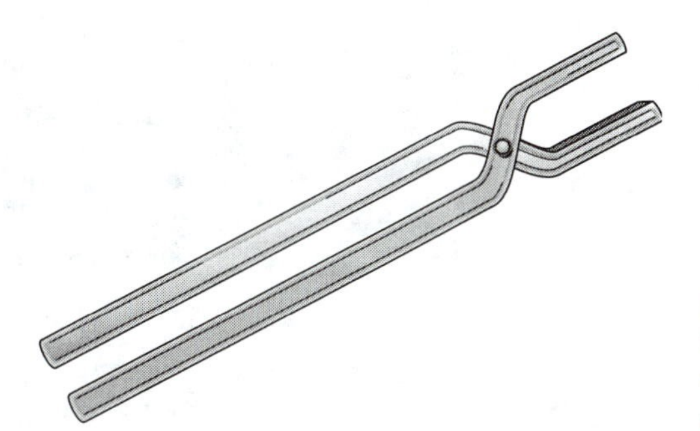

FIGURE 2–21 Using tongs to handle hot metal will extend the life of welding gloves.

Clothing

Wear leathers, cotton, or even wool, but not synthetic fibers that may burn or melt to the skin. Fire retardant fabrics can also be worn. Clothing should cover the skin to shield the body from the arc, which is more intense than sunlight. Ultraviolet light will burn the skin, as will the spatter that welding produces. A leather jacket, leather sleeves, and chaps—although expensive—are preferred. While leather can save clothing from the sparks and spatter of welding, leather does an even better job of protecting the body from burns (Figure 2–22).

Cap

A cap worn under the welding helmet can protect the scalp from the sparks and spatter of welding. A cap also keeps the smoke and dirt in the welding environment from getting into the hair. The bill of the cap to the rear gives protection to the back of the neck.

Tools

Shielded metal arc welding has two indispensable tools for doing quality work. The slag hammer (Figure 2–23) is used to chip away the slag that forms on the **weld bead.** The wire brush is

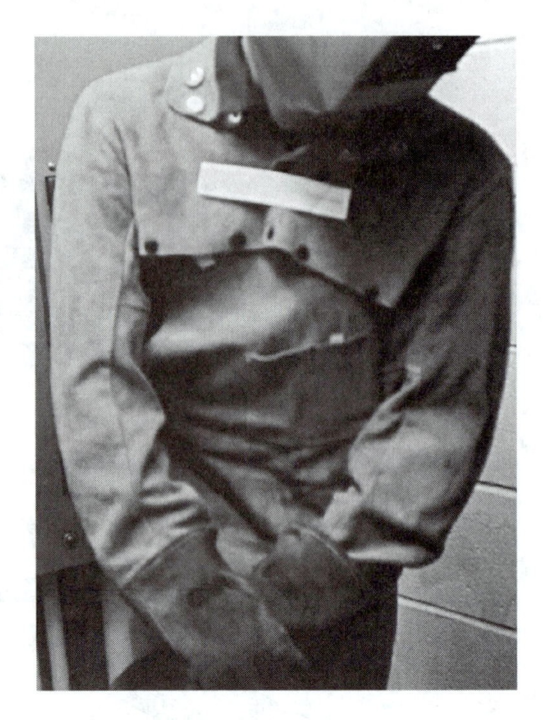

FIGURE 2–22 Leathers protect the clothing and the skin underneath.

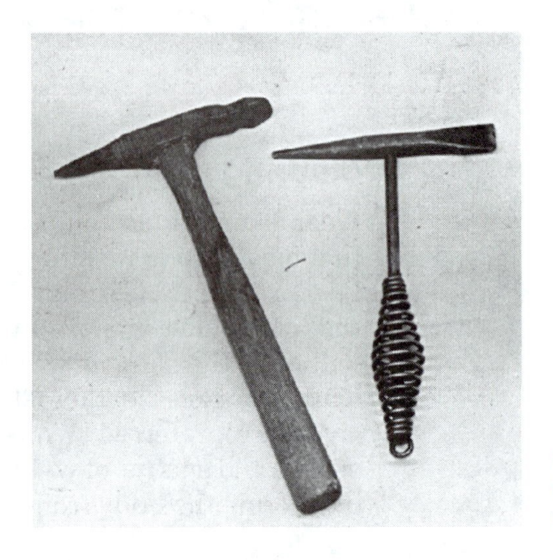

FIGURE 2–23 The slag hammer is a tool of the trade.

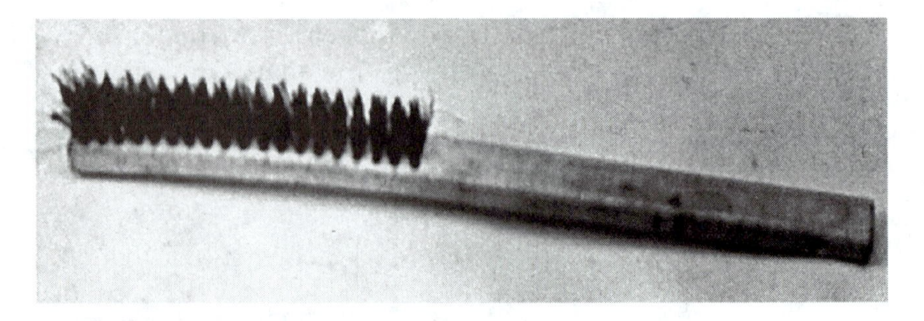

FIGURE 2–24 The wire brush is another essential tool.

FIGURE 2–25 Ventilation systems remove the smoke and the fumes of welding. The Torit is a portable system manufactured by the Donaldson Company of Minneapolis, Minnesota.

used to remove smoke stains and bits of debris (Figure 2–24). Slag left on the weld can be trapped inside other weld beads that follow. Slag trapped inside as weld beads are layered on top of one another affects the quality of the weld.

Ventilation

Smoke and the fumes produced during welding can be eliminated by a ventilation system. A simple blower system can be designed using a small motor attached to a squirrel cage connected to a flexible plastic pipe. Ventilation systems with specially designed filters (Figure 2–25) are available commercially. Potentially harmful substances such as paint, solvents, oils, and galvanized coatings should be removed from the joint surface of the base metal before welding. If this cannot be done, special care must be taken so as not to breath in toxic fumes. Care must always be taken to protect your respiratory system from inhaling the injurious byproducts of welding.

SAFETY REMINDERS

A list of safety reminders for shielded metal arc welding should include the following:

1. Wear safety glasses with side shields for all shop activities.
2. Wear welding gloves with gauntlets to protect hands and lower arms from burns and shocks.
3. Cotton clothing, fire retardant fabrics, and leather offer protection from exposure to the effects of the arc.
4. Be sure the shop is clear of volatile and flammable materials, Figure 2–26.
5. Do not bring cigarette lighters that are under pressure into the welding shop.
6. Always wear a welding helmet with the correct filter lens to protect against arc flash.
7. Make sure the welding area is equipped with the appropriate fire extinguisher, Figure 2–27.
8. Electricity will kill; dry clothing, and gloves and a dry workplace help prevent shocks.
9. Stretch out the workpiece lead and the electrode lead; never wrap around the body.
10. Never weld on pressurized cylinders or on containers that have held volatile substances; never place welding leads near such cylinders or containers.
11. Dispose of electrode stub ends properly to maintain a clean and safe work area, Figure 2–28.
12. Remove the electrode before setting the electrode holder down.
13. Make sure worn **insulators** on the electrode holder and frayed leads are repaired.
14. Make sure any loose connection on the workpiece connection is repaired to maintain electric contact.
15. Use common sense and pay close attention to the work. Avoid daydreaming and distractions.

LAB EXERCISES

1. Become very familiar with the welding helmet you will be using:
   a. Adjust it so that it fits properly on your head.
   b. Examine for cracks in the helmet and the filter lens.
   c. Clean the entire filter lens assembly of smoke stains.
   d. What is the shade number of the filter lens?
   e. Is the shade number sufficient for the welding you will be doing?

2. What system of ventilation will be employed in your shop to keep toxic fumes produced by welding out of the lungs? (Be sure the ventilation system is turned on before beginning any welding.)

FIGURE 2–26 Remove volatile and flammable materials from the welding area. Gasoline, paint thinner, and oily rags can be dangerous anywhere, but especially in the welding area.

FIGURE 2–27 The A-B-C fire extinguisher is an important piece of safety equipment.

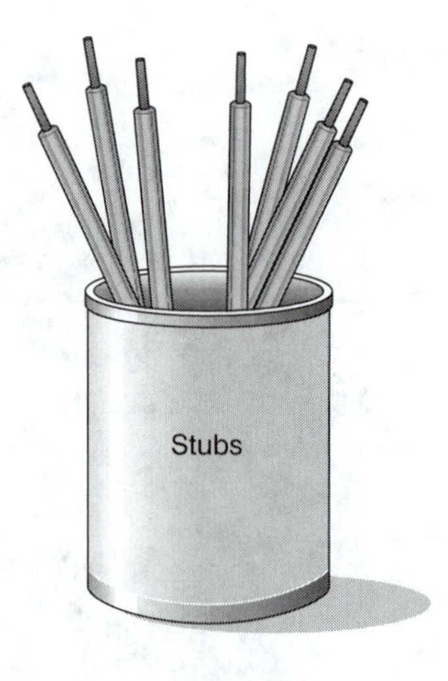

FIGURE 2–28 Properly dispose of electrode stubs so they don't end up causing an accident.

3. What will the welders in your shop be wearing to protect their eyes and hearing?

REVIEW QUESTIONS FOR UNIT 2

Multiple Choice

Choose the best answer to complete the statement.

1. An A-B-C fire extinguisher:
   a. Is for use on electrical fires.
   b. Is for use on volatile substances.
   c. Is for use on combustibles.
   d. All of the above.
   e. None of the above.

2. Choosing a power source should include:
   a. Safety.
   b. Price.
   c. Welding applications.
   d. Input power.
   e. All of the above.

3. Nameplate information on the power source includes:
   a. Manufacturer's name.
   b. Duty cycle.
   c. Class.
   d. Rated load amperage.
   e. All of the above.

4. ____________ means the power source is rated for 6 minutes out of every 10 minutes.
   a. A 30% duty cycle
   b. A 40% duty cycle
   c. A 60% duty cycle
   d. A 10% duty cycle
   e. None of the above

5. A transformer power source:
   a. Is an expensive DC power source.
   b. Uses a rectifier.
   c. Is normally single-phase.
   d. All of the above.
   e. None of the above.
6. Types of power sources include:
   a. Transformer, transformer/rectifier; inverter; plasma.
   b. Transformer, transformer/rectifier; inverter; propane.
   c. Transformer; inverter; diesel; motor generator; solid-state.
   d. Transformer, transformer/rectifier; inverter; engine driven; motor generator.
   e. Transformer; inverter; diesel; gasoline; motor generator.
7. Safety glasses:
   a. Protect the eyes.
   b. Should be worn under the welding helmet.
   c. Must be worn in the shop area.
   d. Must have side-shields.
   e. All of the above.
8. Materials that may require use of a respirator include:
   a. Base metals.
   b. Electrodes.
   c. Objects in the welding area.
   d. All of the above.
   e. None of the above.
9. Shade number ____ is safe for welding with 80 to 174 amperes.
   a. 9
   b. 10
   c. 12
   d. 13
   e. None of the above
10. Clothing worn during welding includes:
    a. leather, cotton, wool.
    b. fire-retardant.
    c. leather, cotton, wool, fire-retardant.
    d. fire-retardant, synthetic.
    e. leather, wool, fire-retardant, synthetic.

Short Answer

1. Name five major safety concerns for shielded metal arc welding.
2. What is the danger from sparks given off by welding?
3. List three things to consider in buying a power source for welding.
4. What is a duty cycle?
5. What is the latest technology in power sources for welding, and what are its advantages?
6. Give two or more reasons for wearing safety glasses.
7. Explain arc flash.
8. Why should you wear welding gloves?

9. Name two volatile subtances.
10. Name two flammable materials.

ADDITIONAL ACTIVITIES

Fill in the Blanks

1. ______________ ______________ will start fires and should be kept out of the welding shop.
2. ______________ is the latest technology in power sources.
3. ______________ provides the minimum electrical information for a power source.
4. ______________ ______________ will eliminate smoke and the fumes produced.
5. Less ______________ of the joint will result in less ______________.
6. One lead is attached to the ______________ ______________ and one lead is attached to the ______________ ______________.
7. ______________ ______________ power source gives the welder more control over the welding parameters.
8. ______________ must be removed before adding another pass.
9. Do not operate shop equipment without wearing both ______________ ______________ and ______________ ______________.

Writing Skills

1. Write an explanation of what to look for in purchasing a power source for welding.
2. Write a paragraph explaining your own understanding of common sense.
3. Based on your reading, write a paragraph on things a welder can do to prevent shocks.

Math

Solve for the percentages:

1. 1/10 = %. **2.** 3/10 = %. **3.** 4/10 = %.

4. 5/10 = %.

5. Welding 6 minutes out of 10 equals _____ %.

6. Welding 2 minutes out of 10 equals _____ %.

Welding power sources are classified on the basis of duty cycle. The duty cycle is the ability to deliver rated output. During welding a power source generates heat. The duty cycle refers to the amount of heat a power source will tolerate without damaging its components at maximum amperage.

Find maximum amperage for 100% (continual) welding. Hint: maximum amperage times the square root of its duty cycle equals the range at which constant welding will not damage the power source.

60% duty cycle: $\sqrt{.60}$ = .774

30% duty cycle: $\sqrt{.30}$ = .547

20% duty cycle: $\sqrt{.20}$ = .448

7. A 400-amp power source with a 60% duty cycle.

8. A 300-amp power source with a 60% duty cycle.

9. A 250-amp power source with a 30% duty cycle.

10. A 180-amp power source with a 20% duty cycle.

An air carbon arc cutting attachment (see Figure 2–29) can be used to both cut metal and gouge out welds. A power source of at least 400 amperes is recommended to assure operating at 100% duty cycle without damaging the power source. This method of cutting metal requires compressed air and carbon electrodes that are used up during the cutting process. Answer the following questions about this type of device.

11. A $\frac{3}{16}$-inch diameter carbon electrode (DCEP) for cutting can operate smoothly at from 150 to 200 amperes. Will a 300-amperage (60% duty cycle) power source operate at 100% duty cycle within the range of a $\frac{3}{16}$ diameter carbon electrode? Show your work.

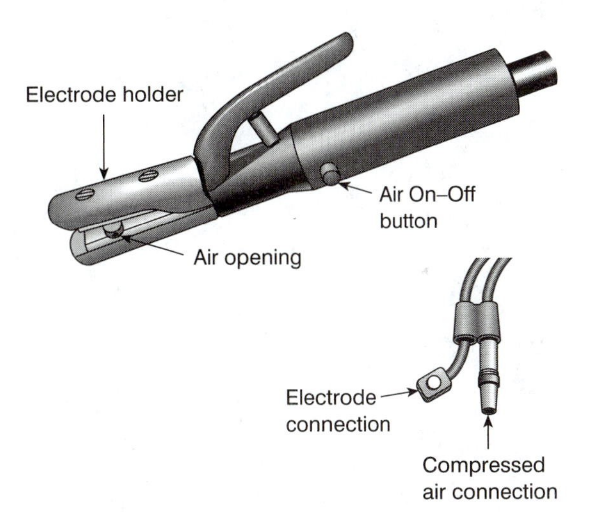

FIGURE 2–29 The air carbon arc torch can be used for gouging out welds. Note that the electrode holder is attached to one connection and compressed air is attached to a second connection.

Carbon electrodes (DCEP)	
Diameter	*Amperage*
5/32 inch	90 to 150
3/16 inch	150 to 200
1/4 inch	200 to 400

12. A $\frac{5}{32}$-inch diameter carbon electrode can operate smoothly at from 90 to 150 amperes. Will a 250-amperage (30% duty cycle) power source operate at 100% duty cycle within the range of a 5/32 diameter carbon electrode? Show your work.

13. A 1/4-inch diameter carbon electrode can operate smoothly at from 200 to 400 amperes. Will a 350-amperage (60% duty cycle) power source operate at 100% duty cycle within the range of a 1/4 diameter carbon electrode? Show your work.

Puzzlers

1. A welder is called into a pit to repair a broken bracket on a piece of equipment. The pit has an inch of water mixed with oil at its bottom. Every time the welder tries to put the electrode into the holder, a shock is received. Can you give any suggestions for completing this job without receiving electrical shocks? Are there other possible safety hazards?
2. The amperage settings on an old welding machine are recognizable, but just barely. After turning the crank and setting the power source, the welder strikes the arc but it continually goes out and the electrode sticks to the plate. After asking the owner if the welding electrodes are about as old as the power source, the owner replies he doesn't know. What can be done to get the power source working?

UNIT 3

Welding Materials

1. METALS FOR WELDING

Properties of Metals

Most metals can be joined together by welding. Metal has many structural shapes with unlimited possibilities for welding ideas, as shown in Figure 3–1. Its **properties** are the characteristics by which a metal is identified. Metal can be light or heavy, hard or soft, to name just a few of its many properties. Metals can also be **ferrous** or **nonferrous.** Ferrous metals contain iron. Ferrous metals include the carbon steels and **cast irons.** Nonferrous metals, which are nonmagnetic, include aluminums and brasses.

The properties of a metal are the results of the ingredients that go into its manufacture. The amount and kinds of ingredients will determine the composition of a metal and how it will respond in a given application. For example, the addition of the element **silicon** to a batch of **steel** may affect its resistance to corrosion. The following is a list of certain properties all metals have, depending upon their added ingredients.

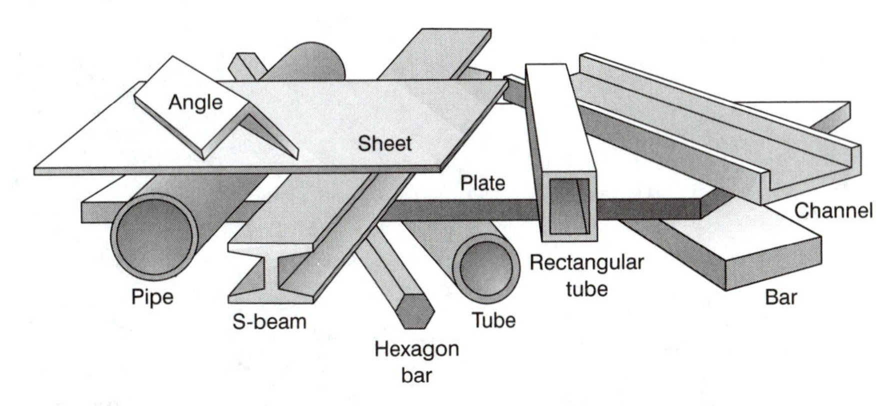

FIGURE 3–1 A variety of structural shapes of metal are required for the many fabrications used in daily living.

Tensile strength: resistance of a material to being pulled apart (see Figure 3–2)

Impact strength: resistance of a material to sudden force without fracture

Shear strength: resistance of a material to two equal forces in the opposite direction

Fatigue strength: resistance of a material to changing forces or loads

Hardness: resistance of a material to penetration or denting

Compression strength: resistance of a material to being crushed (Figure 3–3)

Toughness: resistance of a material to failure from a constant force

Ductility: resistance of a material to **deformation** after stretching, twisting, or bending

FIGURE 3–2 Tensile strength is resistance to being pulled apart.

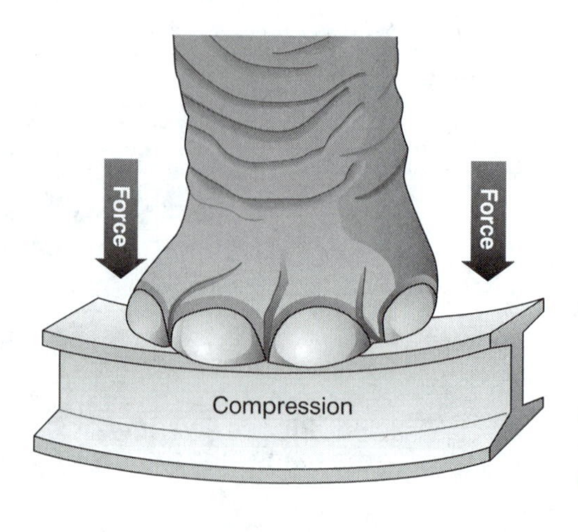

FIGURE 3–3 Compression strength is resistance to being crushed.

Steelmaking

Since all metals are affected by welding, some basic understanding of the behavior of steels during welding is important. With the birth of modern steelmaking in the 1850s, the people of the world have enjoyed a long relationship with steel. Until recently, developments in techniques of fabrication have concentrated on steel because of its dominance over other metals.

The steelmaking process begins when **iron ore,** limestone, and coke are fed into a blast furnace. The process is sketched in Figure 3–4. The type of steel manufactured will depend on the requirements of the customer and, of course, the ingredients that go into the recipe. **Manganese** is an ingredient added to steel to increase its hardness. **Chromium** is an ingredient added to steel to increase its **corrosion resistance. Molybdenum** is a third element sometimes added as an ingredient to steel to increase its tensile strength.

Four general types of steel are classified as (a) **low-alloy steels,** (b) **high-alloy steels,** (c) tool steels, and (d) **carbon steels.** A low-alloy steel, according to the American Iron and Steel Institute, is a steel in which one of the following amounts is exceeded: 1.65 percent manganese content, 0.60 percent silicon content, or 0.60 percent **copper** content. Or, a low-alloy steel is a steel in which a definite minimum quantity of aluminum, **boron,** chromium (up to 3.99 percent), or a minimum quantity of any other alloying element is added to achieve a desired result.

A high-alloy steel is a steel in which chromium, manganese, or nickel content equals 12 percent or better. **Stainless steels** are examples of high-alloy steels because their chromium content exceeds 12 percent. Stainless steels are somewhat more demanding to weld than mild steels because they have (a) a lower melting temperature; (b) lower heat conductivity; and (c) a higher rate of expansion when heated. As a result, twice as many tack welds should be used to reduce distortion. A faster travel speed is suggested with intermittent welding to promote more rapid cooling.

Carbide precipitation occurs when chromium-nickel stainless steels are held within a temperature range of 800° to 1600° for an extended period. Carbon reacts with chromium in a chemical reaction that does not have corrosion resistance.

Tool steels are steels used in cutting or forming applications. The carbon steels in this classification contain from 0.60 to 1.5 percent carbon. The alloy steels in this classification contain in addition to the high carbon, added amounts of chromium, **tungsten,** and molybdenum.

Carbon steels are popular types of steel for welding. The amount of carbon in steel, determined during manufacturing at the steelmaking plant, helps to specify its hardness (Figure 3–5). When the amount of carbon is increased in a batch of steel, the hardness and tensile strength will also increase. However, both the ductility and weldability will decrease.

Steelmaking begins with iron ore, limestone, and coal taken from the earth. This diagram shows some major steps in the process.

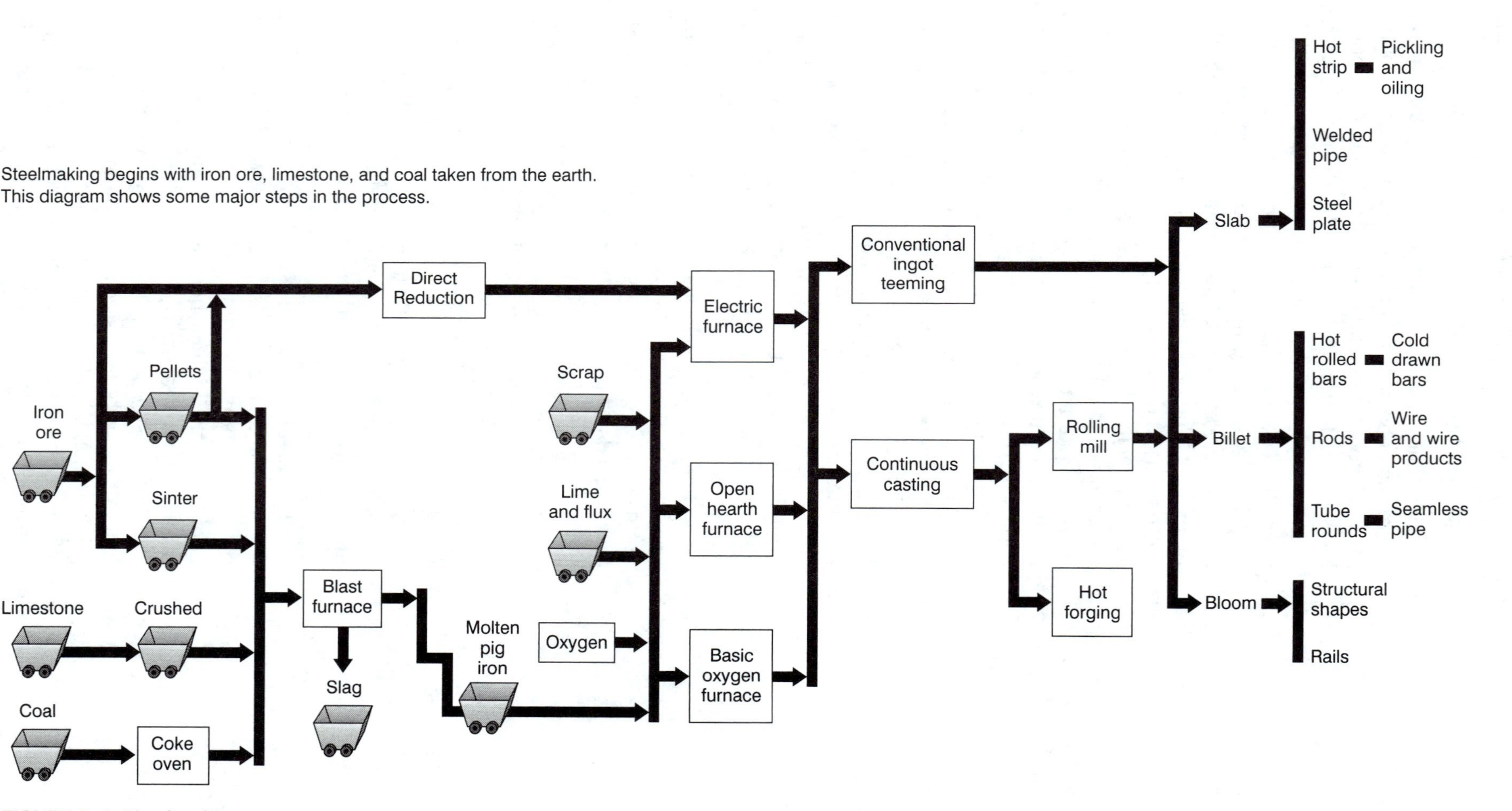

FIGURE 3–4 Steelmaking.

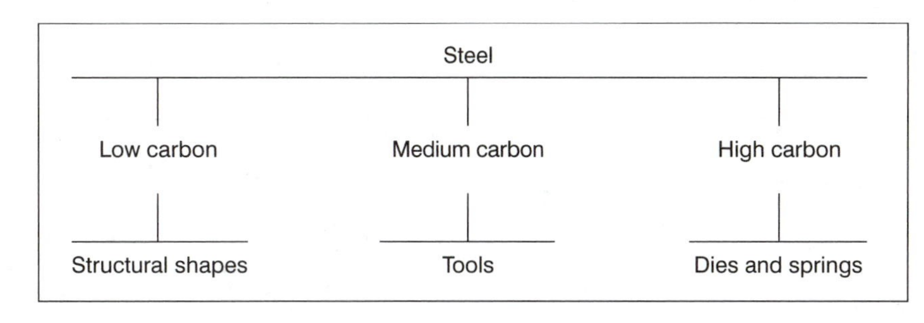

FIGURE 3–5 Some of the uses of different carbon steels.

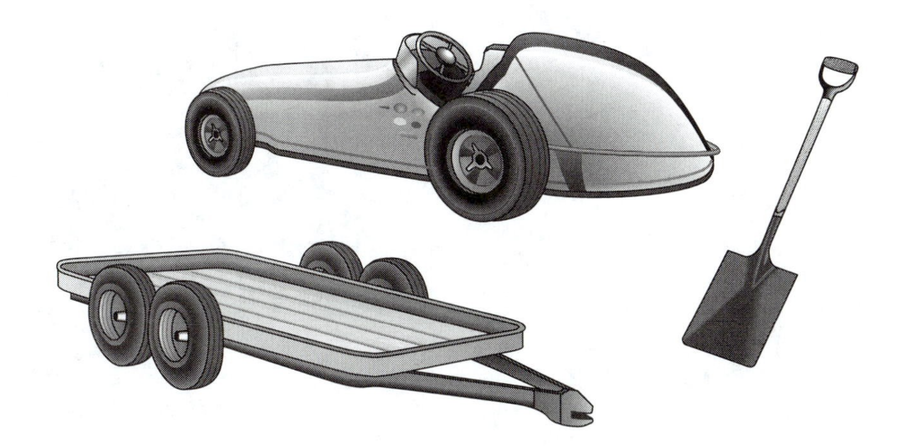

FIGURE 3–6 Varied products made of steel.

Low-Carbon Steel

Low-carbon steel, which is also called **mild steel,** contains from 0.1 to 0.3 percent carbon (this is less than 1 percent). Low-carbon steel is commonly welded, finding application in construction and fabrication. Low-carbon steel cannot be hardened significantly by heat treatment alone. So low-carbon steel has no applications for **tempered** tools such as punches and chisels.

Normally, low-carbon steel does not require additional **preheating** or **postheating** procedures to guard against weld failure. Two exceptions that may require preheating have to do with the outside ambient temperature and the thickness of the base metal. In short, low-carbon steel is easily welded. Low-carbon steels have application in the structural members used for constructing skyscrapers and bridges. Buildings, piping, pressure vessels, and motor frames are general examples of products manufactured from low-carbon steel. Many of the products found around the home are constructed of low-carbon steel (see Figure 3–6).

While low-carbon steel is readily welded, such is not the case with other steels. **Medium-carbon steel** (0.3 to 0.45 percent carbon), **high-carbon steel** (0.45 to 0.65 percent carbon), **very high-carbon steel** (0.65 to 1.5 percent carbon), and cast iron (up to 4.5 percent carbon), all require special care in welding. Table 3–1 shows some of the applications of different steels. Special care

TABLE 3–1 Applications of different steels.

Type of Steel	*Examples of Applications*
Low-carbon	structural shapes (channel, angle, bar)
Medium-carbon	axles, connecting rods, shafts
High-carbon	vehicle springs, band saws, anvils
Very high-carbon	lathe tools, knives, chisels, punches

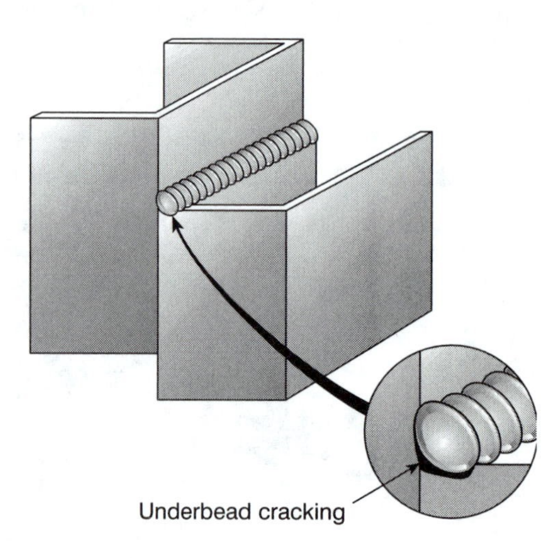

FIGURE 3–7 Underbead cracking results from hydrogen diffused into the heat-affected zone.

may require preheating and/or postheating procedures. High-alloy steels in which elements like chromium are added in a sufficient amount equal to 12 percent or better may require special care.

In working with other types of steel than low-carbon, the welder must know the type of steel being welded. Appearance is not going to help in recognizing it since most carbon steels look alike. Even if low-carbon steel is the most prevalent, a buyer should learn to depend on a reputable supplier for purchasing structural steel. Buying a known classification of steel under the Unified Numbering System from a supplier will prevent cracking of a joint and possible weld failure. The welder should know if a particular steel requires special preheat or postheat treatment to prevent problems such as that shown in Figure 3–7.

ARC FLASH

All metals have different melting points:

Steel (0.4–0.7% carbon) 2500°

Stainless steel (nickel) 2550°

Steel (0.15–0.4% carbon) 2600°

Stainless steel (low-carbon) 2640°

Steel (less than 0.15% carbon) 2700°

Identifying Low-Carbon Steel Low-carbon steel cannot be distinguished from the other types of carbon steels by using a magnet. However, a magnet will isolate the carbon steels from other metals that are nonmagnetic such as copper, brass, and aluminum.

Low-carbon steel can sometimes be identified from other types of carbon steel by filing or by use of a chisel. A file made for use on steel or a chisel made for use on steel will bite easily into low-carbon and medium-carbon steel to remove metal chips. See Table 3–2.

Apply pressure with the file to the unknown steel on the forward stroke only, as illustrated by Figure 3–8. The file and the

TABLE 3–2 Recognizing metals by filing or chiseling.

Base Metal	*Use of Chisel*	*Use of File*
copper	chips easily	bites easily
brass	chips easily	bites easily
aluminum	chips easily	bites easily
low-carbon steel	continuous chip	bites easily
medium-carbon steel	chips easily	bites with pressure
high-carbon steel	difficult to chip	does not bite

FIGURE 3–8 Use of a file to aid in determination of carbon content.

chisel will remove chips either with difficulty or not at all as the percentage of carbon content increases into high-carbon and very high-carbon steel.

REVIEW

1. Why is low-carbon steel more easily welded than other types of steel?
2. How does the addition of carbon affect steel?

Lab Exercises

1. Take pieces of steel supplied by your instructor and, using a file or a chisel, attempt to identify the types of steel.
2. Take a wire supplied by your instructor, and bend it several times:
   a. Describe the effects of bending.
   b. Which of the properties of metals have you been applying in your examination?

2. STEEL PLATE

Preparation for Welding

A structural quality low-carbon steel will be needed for the exercises in this book. According to standards established by the American Society for Testing and Materials (ASTM), A36, A283, A570 are three steels suitable for welded structures. The recommendation is for a general grade structural steel that can be readily welded without added preparation beyond those laid out in this unit.

Plate is one structural metal shape. Plate is defined as being over 3/16 inch thick and over 6 inches wide. **Sheet** is another structural metal shape. Sheet is 3/16 inch thick and less, but over 6 inches wide. A thickness of 1/4-inch to 3/8-inch thick steel plate is recommended for all exercises in this text (see Unit 6 and the Appendix). The thinner the material, the more distortion will be experienced. The padding exercise requires plate that is 3/8-inch thick, and 6 inches wide by 8 inches in length (Figure 3-9).

Flat bar is another structural metal shape. Flat bar is over 3/16 inch thick and up to 6 inches wide, commonly available in 20 foot lengths. Several lengths 8 inches long should be cut out of 3 inch flat bar for the corner joint and tee joint exercises (Figure 3–10). Angle iron, 3-inch, can be used as a substitute for flat bar in the corner joint exercise.

The first step in preparing for welding exercises is to cut steel to the dimensions specified. Flat bar (over 3/16 inch thick) is easier to work with coming in 6 inch widths that can be cut 8 inches long. Unlike plate, flat bar requires a single cut to prepare for the welding exercises presented in this book. Some metal suppliers will cut material to size as requested. With training, you can learn to use an **oxyacetylene torch,** air carbon arc, or a band saw to do the job (see Figures 3–11 and 3–12).

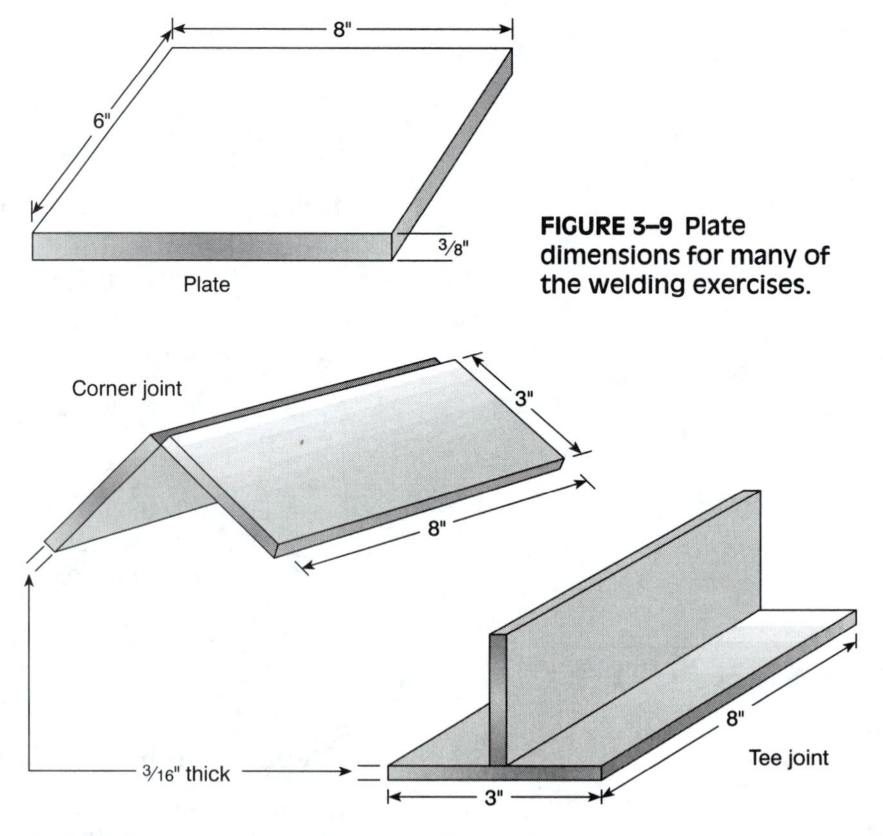

FIGURE 3–9 Plate dimensions for many of the welding exercises.

FIGURE 3–10 Corner joint and tee joint dimensions for welding exercises.

FIGURE 3–11 Torch designed for oxyacetylene cutting.

FIGURE 3–12 The horizontal band saw is an important tool for metal cutting. It can be used to cut coupons for weld testing.

FIGURE 3–13 The grinder is an important shop tool, used to clean metal surfaces for grinding or beveling the edges.

Mill scale and any **rust** should be removed from the surfaces of the joint where the welding will be completed. A hand held grinder (Figure 3–13) is a useful tool for this purpose and belongs in any shop where welding is done. A versatile tool, the grinder can be used for quickly grinding away welds. Grinding adds to the appearance of any finished fabrication project.

REVIEW

1. How is mill scale different from rust?
2. How is mill scale like rust?

Lab Exercises

1. Make a list of the ingredients and/or properties of grades of steel used in the shop.
2. Make a list of the different structural shapes for metal you are able to identify in the shop.
3. Based on information provided by the instructor, write the safety precautions to be followed for the tool(s) used in cutting metal plates to size for welding.

3. WELDING ELECTRODES

Composition and Identification

Shielded metal arc welding requires electrodes (Figure 3–14). The electrode is the filler metal added to the joint. Electrodes must be matched to the base metal being welded. For example, carbon steel electrodes are designed for welding carbon steels of the re-

FIGURE 3–14 Electrodes are the filler metal.

quired tensile strengths. The properties of the weld can be affected by many factors including the choice of the electrode. Product manufacturers must design welding procedures to test the effects a given electrode will have on the weld.

Each electrode has a flux covering with a solid wire core. The flux burns during the welding process and produces a gaseous shield that protects the weld from the air. The flux also determines penetration into the base metal and helps to stabilize the arc to make welding easier. The lighter flux rises to the surface of the weld pool created by the arc, forming slag. Slag that protects the weld bead as it cools must be chipped off with a slag hammer before laying another weld bead.

Electrodes are identified by the numbering classification system developed by the American Welding Society. Through use of this numbering system, a welder does not have to rely on brand names when trying to match the electrode to the base metal. A typical electrode number might be:

E6011

The **E** in the number means an arc welding electrode. The first two digits, **60** (multiplied by 1000 = 60,000), refer to the minimum **tensile strength** of the steel filler metal per square inch. The third digit, **1,** provides information on the selection of **welding positions** in which the electrode can be used. A number 1 means the electrode can be used in all four welding positions. While the emphasis in Unit 6 will be placed on exercises to be completed in the **flat position** and the **horizontal position,** the Appendix contains exercises for the **vertical position** and the **overhead position.** The Appendix lists exercises of the type used to pass qualification tests developed by the American Welding Society according to the *Structural Welding Code—Steel D1.1.*

The fourth digit, **1,** gives the electric current selection. A number 1 means electrodes are available using AC, DCEN and DCEP, that is, all three current choices. The fourth digit also provides information on the flux. A number 1 means a cellulose potassium flux. A cellulose flux is affected by excessive heat. For example, E6011 should be kept in a dry place, but not in a heated oven. More detailed information on electrodes can be found in Unit 8.

SAFETY REMINDER

Electrode stub ends should be placed immediately in a container to prevent the possibility of a fire or other hazards such as a tire puncture.

REVIEW

1. What is a welding electrode?
2. What is the purpose of the flux covering?
3. What is a purpose for having a numbering classification system to identify electrodes?

4. PARTS OF THE WELDING CIRCUIT

Electrode Holder

The electrode holder is connected by the electrode lead to the power source. The electrode lead is a wire cable. The electrode holder (Figure 3–15) is a device for holding the electrode firmly in position during welding. The welding circuit is completed by the union of the power source to the base metal through a connection with the electrode holder.

A high quality holder can be easily repaired by replacing worn insulators. Insulators are devices made of materials that do not conduct electricity. Worn insulators or a loose or frayed electrode lead must be replaced to prevent shocks or arcing of the electrode holder off the base metal. This equipment must be kept repaired for your own safety.

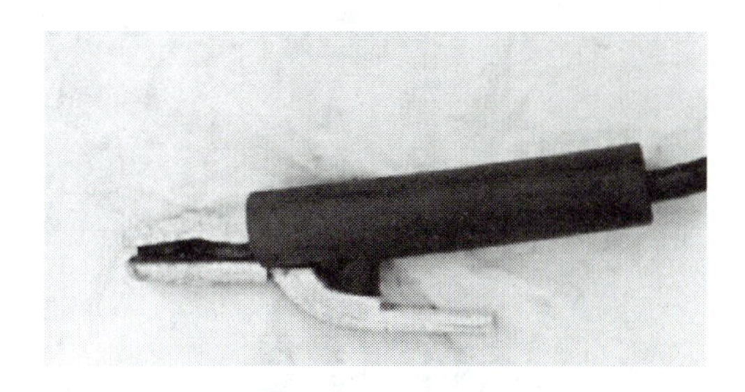

FIGURE 3–15 Electrode holder.

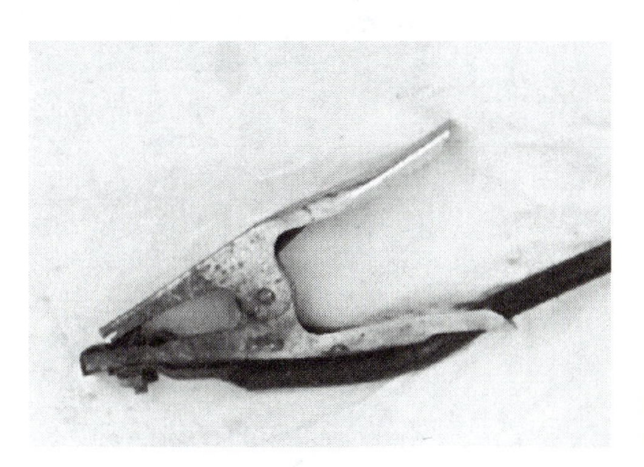

FIGURE 3–16 Workpiece connection.

Workpiece Connection

The workpiece connection (Figure 3–16) joins the base metal to the power source. If there is a problem striking an arc, the workpiece connection should be examined immediately. A bad connection because of rust, scale, or paint on the base metal can prevent the making of an electric circuit. The cause can also be a loose fitting connection at a point where the **workpiece lead** is attached to the workpiece connection.

The workpiece connection should be attached to the base metal as close to the weld as possible. If a connection is not made directly to the base metal at the point of the welding, undesirable arcing may occur, which can damage another part. For example, improper placement of the workpiece connection could damage a hydraulic cylinder, as shown in Figure 3–17. Special care must always be taken in the placement of the workpiece connection because electricity will find the path of least resistance. Be sure you know the path electricity will travel to directly reach the point of the welding.

SAFETY REMINDER

A welder must keep clothing dry and avoid welding while standing on a wet surface, so as to avoid injury from an electric shock. Clothing that is wet from perspiration can provide enough moisture to cause violent muscular contractions.

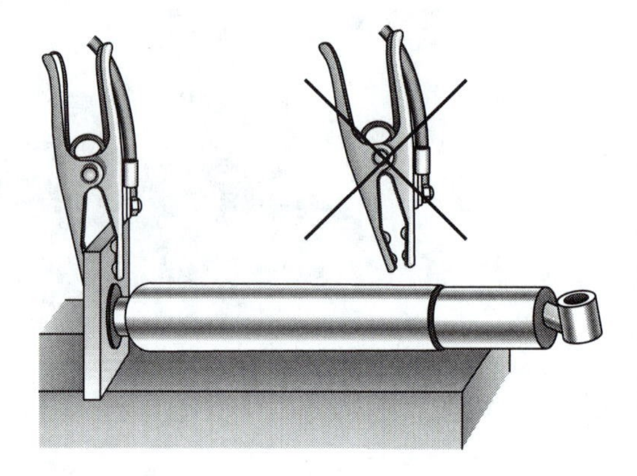

FIGURE 3–17 Placement of the workpiece connection can be important to avoid damage to equipment. In this example it is important to keep the current from passing through the hydraulic cylinder and arcing so as to cause damage to the cylinder wall.

REVIEW

1. Does it make any difference where the workpiece connection is placed before welding?
2. What would cause the electrode holder to arc on to the base metal during welding?
3. What does it tell the welder if the electrode holder is unable to hold the electrode firmly in position during welding?

Lab Exercise

1. Examine the condition of the electrode holder and the workpiece connection to be used for welding in your shop.

REVIEW QUESTIONS FOR UNIT 3

Multiple Choice

Choose the best answer.

1. Ferrous metals:
   a. Are nonmagnetic.
   b. Are aluminums.
   c. Are brasses.
   d. Are carbon steels.
   e. All of the above.
2. Fatigue strength is:
   a. Resistance to sudden force without fracture.
   b. Resistance to penetration.
   c. Resistance to failure from a constant force.
   d. Resistance to changing forces.
   e. None of the above.
3. Toughness is:
   a. Resistance to deformation after twisting.
   b. Resistance to being crushed.
   c. Resistance to sudden force without fracture.
   d. Resistance to being pulled apart.
   e. None of the above.

4. Shear strength is:
   a. Resistance to being pulled apart.
   b. Resistance to denting.
   c. Resistance to being crushed.
   d. Resistance to sudden force without fracture.
   e. None of the above.

5. Stainless steels:
   a. Are steels in which chromium content exceeds 12 percent.
   b. Are high-alloy steels.
   c. Are easier to weld than mild steels.
   d. b and c.
   e. a and b.

6. _____________ are steels in which 1.65 percent manganese content is exceeded.
   a. High-alloy steels
   b. Carbon steels
   c. Low-alloy steels
   d. Tool steels
   e. None of the above

7. Carbon affects the:
   a. Hardness of steel.
   b. Ductility of steel.
   c. Strength of steel.
   d. All of the above.
   e. None of the above.

8. _____________ does not usually require special care in welding.
   a. High-carbon steel
   b. Medium-carbon steel
   c. Low-carbon steel
   d. Very high-carbon steel
   e. High-alloy steel

9. _____________ is magnetic.
   a. Copper
   b. Brass
   c. Aluminum
   d. Chromium
   e. None of the above

Short Answer

1. How does toughness differ from hardness?

2. How does impact strength differ from compression strength?

3. Name one reason why low-carbon steel is so commonly used in construction and fabrication.

4. What is the difference between plate and sheet metal?

5. What are two purposes of the flux covering on the electrode?

6. Of what benefit is the slag covering the weld bead?

7. What is tensile strength?

8. What does the third digit of E60**1**1 mean?

9. Name the four welding positions.

10. Give a reason for adding manganese, molybdenum, and chromium in the steelmaking process.

ADDITIONAL ACTIVITIES

Crossword Puzzle

Across

1. Helps determine the hardness of steel
2. Low-carbon steel
3. Metal 3⁄16-inch or less in thickness
4. Added to steel for corrosion resistance
5. Resistance to deformation after stretching, twisting, or bending
6. Application of heat to the base metal before welding
7. Connects electrode holder to power source

Down

1. Nonmagnetic
2. Filler metal
3. Not a drink, but fed into blast furnace as fuel
4. Over 3⁄16-inch thick
5. Liquid metal formed at the point of welding (2 words)

Writing Skills

1. Use the following words in sentences forming a paragraph: *shielded metal arc welding, carbon steel, aluminum, lightweight, base metal, oxide, rust.*
2. Explain in a paragraph the method you will use to prepare a piece of plate for the padding plate exercise.
3. Explain how the amount of carbon affects the welding procedure.

Math

The following symbols can be used to express inequalities: > means one quantity is greater than another, whereas < expresses that a quantity is less than another. Is the number 1 greater than or less than 0.1? You can easily see that 1 > (is greater than) 0.1. Is 0.1% greater or less than 0.1? Changing 0.1% to decimal form (0.001) shows that 0.1% < (is less than) 0.1.

Solve for greater than or less than.

1. 1% __ 0.1 **2.** 0.3% __ 1% **3.** 0.2 __ 0.45% **4.** 0.3% __ 0.45%
5. 0.65% __ 0.65 **6.** 0.65% __ 0.45% **7.** 0.01 __ 1.0%
8. 0.65% __ 1.5% **9.** 4.5% __ 0.045 **10.** 1% __ 0.045
11. 0.45% carbon __ 0.65% carbon
12. 0.65% carbon __ 1.5% carbon
13. 1.5% carbon __ 4.5% carbon

Puzzlers

1. Adding carbon to steel will increase its hardness; how is its ductility and tensile strength affected?
2. If hardness increases by the addition of some element to a batch of steel, will the strength of all its properties continue to improve? Explain. Which is of greater hardness, steel or glass? Explain.

UNIT 4

Fundamentals of Welding

1. ACT OF WELDING

Striking the Arc

Every weld begins by striking an arc. While striking an arc is easier using DC, with persistence striking an arc in AC will become almost as easy. The **scratch method** and the **tap method** are the two ways of striking an arc (Figure 4–1). With the scratch method the electrode is scraped on the base metal until an arc is established, then raised slightly. With the tap method the electrode is placed on the base metal at a single spot and then, by the use of wrist action,

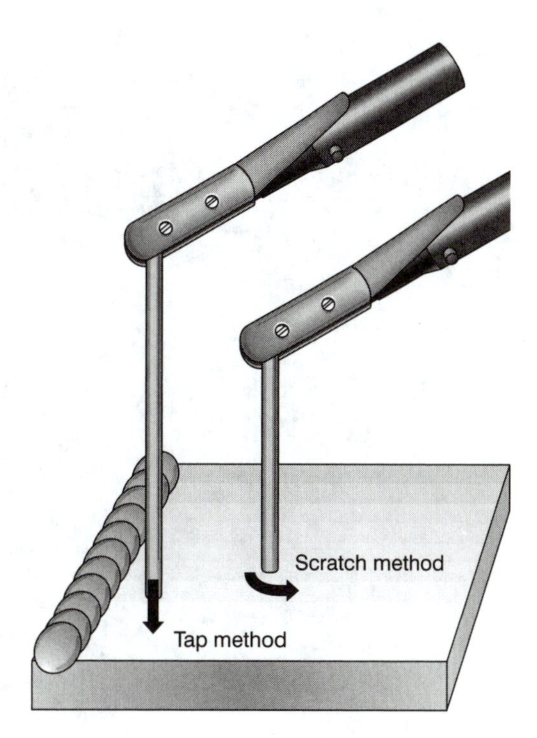

FIGURE 4–1 Two methods for starting an arc.

raised quickly when the arc begins. The tap method is used by skilled welders to keep arc strikes only in the welding area.

Note the position of the electrode in the electrode holder in Figure 4–2. Place the electrode in the electrode holder at a 135° angle for all the exercises in this text. The welder can then burn each electrode down to the stub end with very little waste. Using a 45° or 90° angle will cause some waste.

Care must be taken to see that each **arc strike** is placed inside the welding area. Stray arc strikes outside of the weld produce brittle areas subject to cracking under stress. A careless arc strike can damage parts that were not originally the focus of the welding (Figure 4–3). An instructor will most certainly reject any **weldment** with several arc strikes outside of the welding area.

Tying weld beads end to end to each other is important. Put the arc strike in the **crater** of the previous weld bead and continue to lay the next weld bead from that spot, as shown in Figure 4–4. The practice of keeping arc strikes in the welding area will contribute to quality welds that meet recognized standards.

Welding Variables

Variables such as the size and variety of electrode, the kind and thickness of the base metal, the type of joint, and the position for welding establish the **parameters** for selecting the choice of current and the amperage setting. Such welding techniques as the arc length, travel angle, work angle, and travel speed are affected by the parameters that are established.

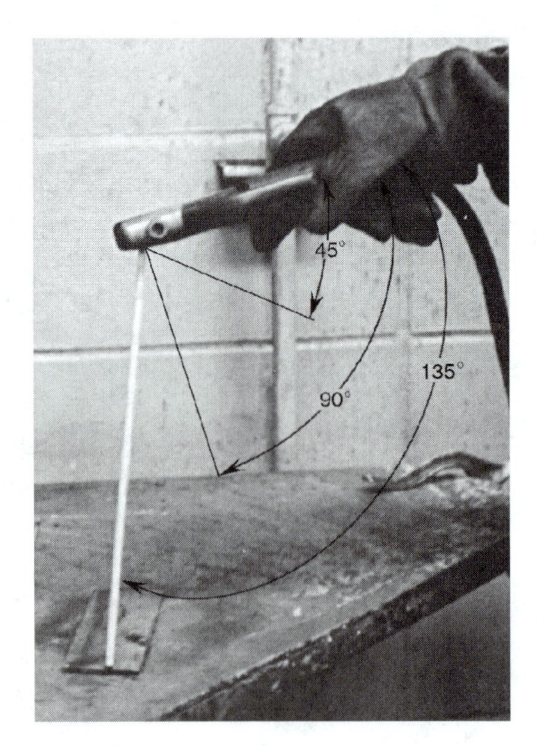

FIGURE 4–2 Electrode positioned at a 135° angle within the holder will prevent waste.

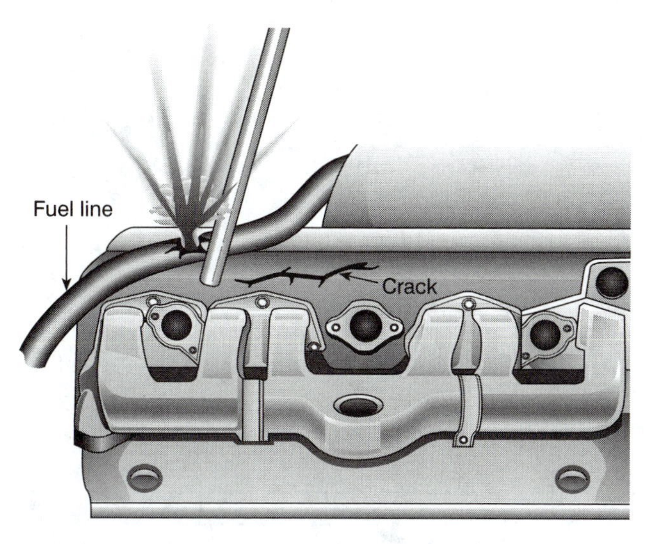

FIGURE 4–3 Placement of the electrode striking an arc can be important to avoid damage to equipment or even worse consequences.

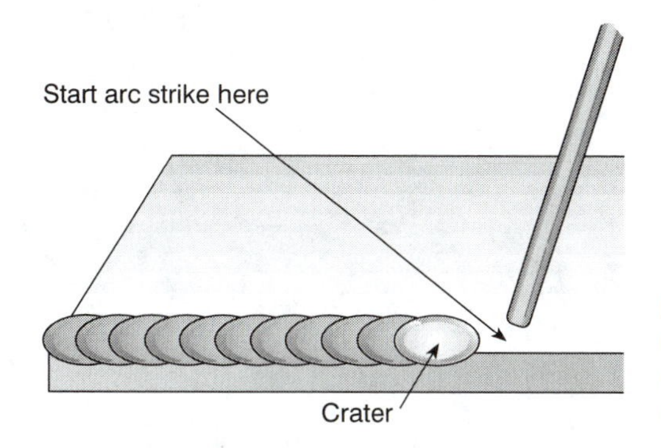

FIGURE 4–4 Position the next electrode within the crater of the previous weld bead.

Arc Length The **arc length** is the distance of the electrode from the weld pool. This distance is equal to approximately the size (the diameter) of the electrode. A measurement of diameter does not include the flux covering, which will vary with the type of electrode. For example, E6011 has a thinner flux covering than the thicker fluxed E7024, which contains iron powder in the flux. If the arc length is too long, then the voltage is too high, causing lots of spatter, as in Figure 4–5. This spatter results in balls of metal forming, which produce shallow root penetration of the weld into the base metal. With too long an arc length, the arc becomes unstable, producing a wide bead that provides poor protection from contaminants in the air. If the arc length is too short, then the voltage is too low and the weld bead is narrow with shallow penetration that may cause the electrode to stick to the base metal, (Figure 4–6). With too short an arc length, the arc cannot generate enough heat to melt the electrode and the base metal adequately.

The welder can control the arc voltage and the amperage to some degree by raising and lowering the electrode above the

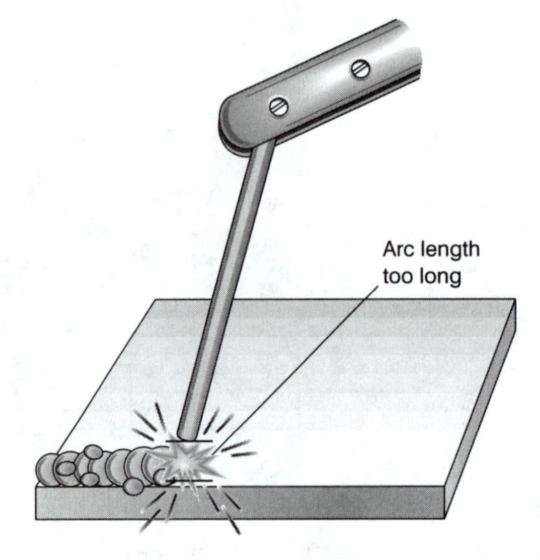

FIGURE 4–5 An arc length that is too long causes spatter.

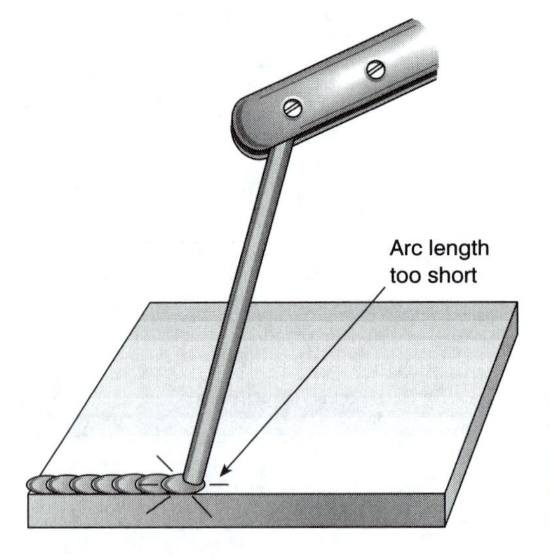

FIGURE 4–6 An arc length that is too short can cause sticking of the electrode to the base metal.

base metal. During welding the arc voltage will vary slightly depending on the length of the arc, as pictured in Figure 4–7. The welder can tell the correct length of the arc during welding by sound. If the arc length is correct, the welding should sound like bacon frying.

Travel Angle There are two angles important to the welder in the act of welding. The **travel angle** is one of these two angles of importance. Travel angle is the angle of the electrode in relation to the direction it is moving along the joint. The electrode should be positioned for welding in the flat position with a travel angle of 10° to 15° from the vertical, as in Figure 4–8. The travel angle will be referred to throughout the welding exercises of this book. Maintaining the correct travel angle will help to keep slag forming behind the weld pool. The welder wants to avoid having the weld pool forming ahead of the electrode, which can result in trapping

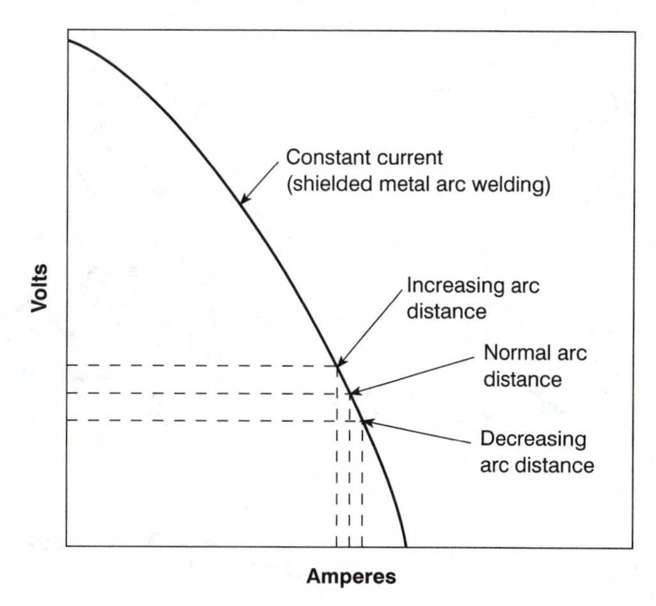

FIGURE 4–7 Change in voltage is matched to arc length. A decrease in arc length results in a decrease in arc voltage. An increase in arc length results in an increase in arc voltage.

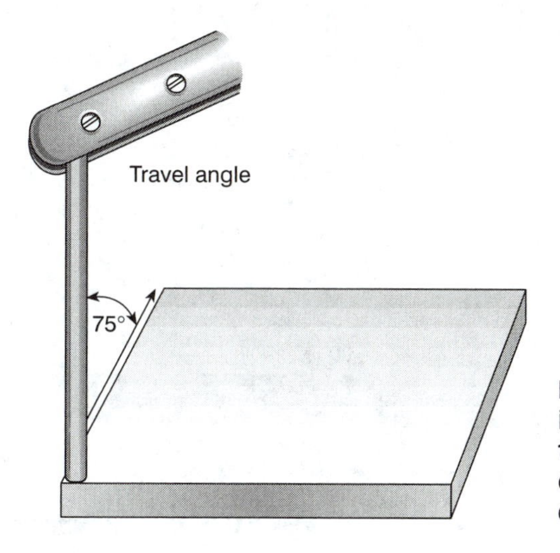

FIGURE 4–8 Travel angle is important for welding technique. Tilt the electrode 15° in direction of travel.

slag in the weld. Trapped slag will affect the weld quality, as shown in Figure 4–9.

Work Angle The **work angle** is the second of two angles important to the welder in the act of welding. The work angle is the angle of the electrode in relation to the base metal. This work angle will vary depending on the job at hand. Welding on plate in the flat position, the 180° angle is bisected with a 90° work angle, as in Figure 4–10. Maintaining the correct work angle will help to avoid **undercut** along the weld, pictured in Figure 4–11.

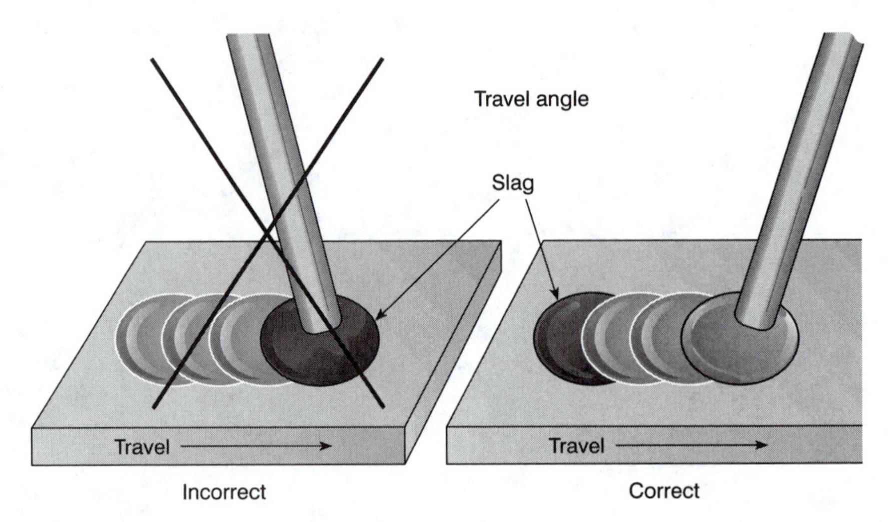

FIGURE 4–9 Avoid trapping slag by maintaining the correct travel angle.

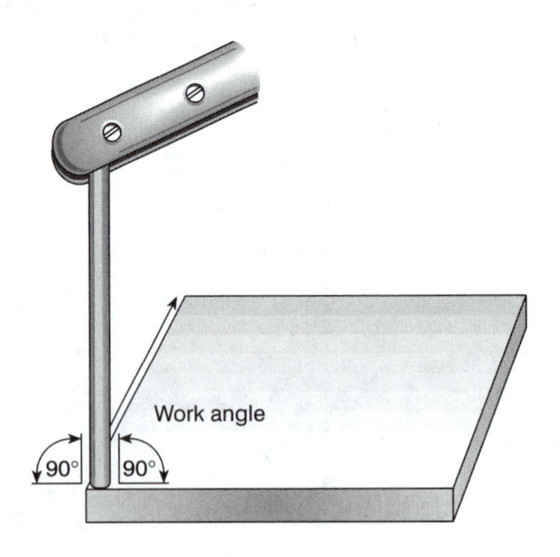

FIGURE 4–10 Work angle is important for welding technique. Note the 90° angle.

Travel Speed The **travel speed** is the movement of the weld pool across the joint. If the travel speed is too slow, the deposited filler metal piles up. Too slow a travel speed makes the weld bead too high and too wide. If the travel speed is too fast, the weld bead is narrow. Too fast a travel speed also causes incomplete **fusion** of the weld with the joint. The tendency is for beginning welders to move too quickly across the joint. Try to keep the width of the weld bead to three times the diameter of the electrode measured by its core wire, not including the flux covering.

ARC FLASH

In the act of welding, pay attention to the position of the weld pool in relationship to the joint.

SAFETY REMINDER

Keep hands and feet dry to avoid shocks. If it becomes necessary to work in a wet environment, stand on a dry insulated platform like a piece of wood.

Current Setting

Many welding problems can be eliminated from the start by learning how to set the amperage on the power source correctly. Once the current has been matched (AC, DCEP, or DCEN) with the welding electrode, setting the amperage should be easy. Use Table 4–1 as a starting point for setting the amperage. If the electrodes are constantly sticking to the base metal, turn up the amperage. If undercut occurs along the weld bead, if the bead spreads out wider than three diameters, or if there is a lot of spatter, turn down the amperage. Begin by choosing a mid-range amperage setting, making adjustments as necessary.

Some power sources give the welder the advantage of adjusting not only the amperage, but also the voltage. Power sources for shielded metal arc welding that offer this opportunity to set both the amperage and the voltage give the welder more control over the arc.

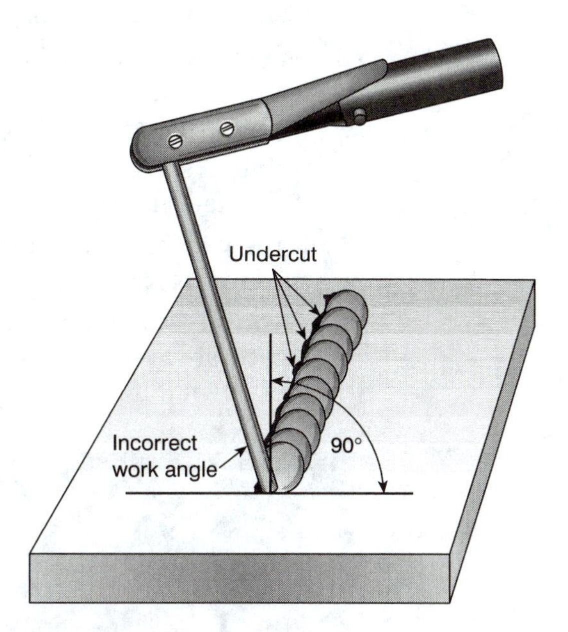

FIGURE 4–11 Undercut is one of the problems caused by an incorrect work angle.

TABLE 4–1 Range of settings.

Diameter of Electrode (inches)	*Amperage*	*Voltage*
1/16	20–40	17–20
5/64	25–60	17–21
3/32	40–120	17–21
1/8	55–170	18–22
5/32	80–270	18–22
3/16	140–330	20–24
7/32	200–400	20–25
1/4	250–500	20–26
5/16	275–525	22–27

After gaining some confidence by using the table, learn to select the amperage setting by the appearance of the weld bead without referring to the table. Remember that there is an acceptable range, so it is not necessary to locate some exact amperage setting.

If the power source has both AC and DC capabilities, a welder will probably use DC most of the time. DCEP provides a more stable arc with greater root penetration than either DCEN or AC. AC gives better penetration than DCEN, with DCEN preferred on thin metal when **melt-thru** could be a problem.

Suggestions for Improving Welding Technique

Remember the following tips to improve welding skills:

1. Always make sure there is good electric contact at the workpiece connection, as close to the welding as possible.
2. Relax and find a comfortable position in which to weld. Quality welding depends on staying loose, which reduces fatigue.
3. Stay aware of both the work angle and the travel angle in the act of welding.
4. Keep both hands on the electrode holder in the early stages of training (see Figure 4–12). Use one hand supporting the other, if that feels more comfortable.
5. If the weight of the electrode lead is causing undo stress, lay the lead over an object in the area.
6. Always remove the slag before completing another pass. For safety, always chip slag away from the body, or keep the helmet down if it is designed with a flip-up filter lens.

FIGURE 4–12 Position both hands on the electrode holder, or grab the forearm of the hand clutching the electrode holder.

REVIEW

1. What are the results of too long an arc length?
2. What is the difference between the work angle and the travel angle?
3. Connect the problem with its cause:

a. high weld bead that is too wide	travel speed too fast
b. narrow weld bead with incomplete fusion	arc length too long
c. spatter with shallow root penetration	amperage too high
d. shallow root penetration with a sticking electrode	travel speed too slow
e. undercut and a wide weld bead with lots of spatter	arc length too short

4. How do you adjust the amperage setting for welding with confidence?

REVIEW QUESTIONS FOR UNIT 4

Multiple Choice

Choose the best answer.

1. The tap method of striking an arc:
   a. Produces brittle welds.
   b. Puts arc strikes in the welding area.
   c. Is the same as the scratch method.
   d. Has the welder scraped the electrode on the base metal.
   e. None of the above.
2. Parameters for selecting a choice of current include:
   a. Size of the electrode.
   b. Kind and thickness of the base metal.
   c. Type of joint and position for welding.
   d. Arc length and travel speed.
   e. a, b, and c.
   f. All of the above.
3. If the arc length is too long:
   a. The arc becomes unstable.
   b. Voltage is too low.
   c. The weld bead is too narrow.
   d. There is only a small amount of spatter.
   e. All of the above.
4. The welder can control arc voltage through:
   a. The angle of the electrode in its holder.
   b. The thickness of the base metal.
   c. Raising and lowering the electrode.
   d. None of the above.
   e. All of the above.

5. If the travel speed is too slow:
   a. Deposited metal piles up.
   b. The weld bead is too wide.
   c. There will be incomplete fusion.
   d. a and c.
   e. a and b.
6. To set the amperage correctly:
   a. Match the current with the welding electrode.
   b. Choose a mid-range to begin.
   c. If electrodes are constantly sticking, turn up the amperage.
   d. If there is a lot of spatter, turn down the amperage.
   e. All of the above.
7. To improve welding skills:
   a. Keep relaxed in a comfortable position.
   b. Stay aware of both the work angle and the travel angle.
   c. Turn the power source off when leaving the shop.
   d. a and c.
   e. a and b.
8. Welding technique includes:
   a. Arc length.
   b. Travel angle.
   c. Choice of electrodes.
   d. Travel speed.
   e. a, b, and d.
   f. a, b, and c.
9. DCEN:
   a. Gives better penetration than DCEP.
   b. Makes for a less stable arc.
   c. Is used most of the time.
   d. Is preferred on thin metal.
   e. Gives better penetration than AC.
10. The work angle:
   a. Is the same as the travel angle.
   b. Can help avoid undercut.
   c. is 10° to 15°.
   d. a and b.
   e. None of the above.

Short Answer

1. Why is the tap method preferred for starting an arc?
2. What angle in positioning electrode to holder can help to eliminate wasting electrodes?
3. What are some of the problems from holding too short an arc length?
4. What are some of the problems from holding too long an arc length?
5. What is the appearance of the weld bead when the travel speed is too fast?
6. What is the appearance of the weld bead when the travel speed is too slow?

7. What does it tell you when the electrode continues to stick to the plate?
8. What does it tell you when the weld bead spreads out with lots of unnecessary spatter?
9. Why try to be relaxed for welding?
10. Where is the workpiece connection placed for welding?

ADDITIONAL ACTIVITIES

Word Find

Circle the technical words in the puzzle. (Words can be found forward, backward, down, up, and diagonally.)

```
J P C L O O P D L E W W T
O R R T L S R T N O E E R
I X A N E N T I R O O L A
T P O I C G T R N I D D V
W W L O P A A G E R T M E
O L E J A L K R S S E E L
R N C L Z S A L E E S N A
K O R O D R E N D P P T N
A I A B G E E N E A M P G
N S T O M S R I S Y M A L
G U E R S P A T T E R P E
L F R T U C R E D N U G X
E D O R T C E L E N U O B
```

TRAVEL ANGLE
ARC
SLAG
WELDER
WORK ANGLE
ELECTRODE
SPATTER
WELDMENT
FUSION
STRESS
UNDERCUT
CRATER
AMPERAGE
WELD POOL
JOINT

Writing Skills

1. Write a paragraph explaining some of the welding techniques.
2. Write an explanation for some of the problems in careless arc strikes.
3. Write the instructions explaining how to use a welding power source from the shop.

Math: Degrees

Look at the sketches depicting common angles and the number of degrees (°) in them.

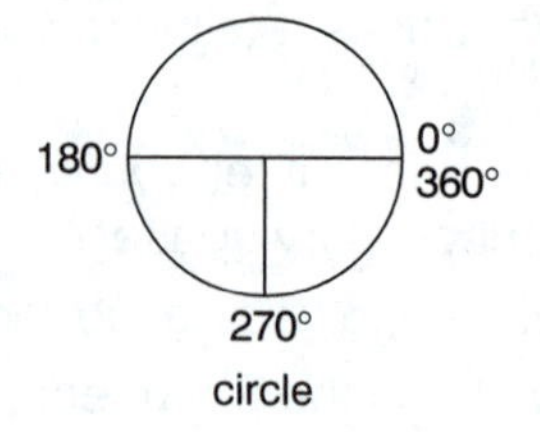

circle

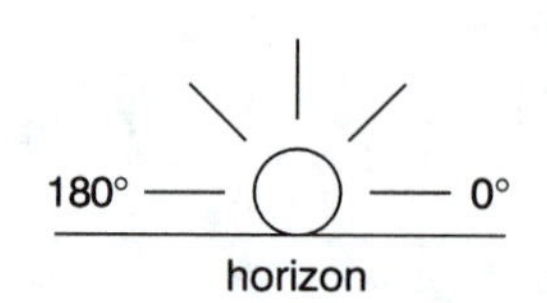

horizon

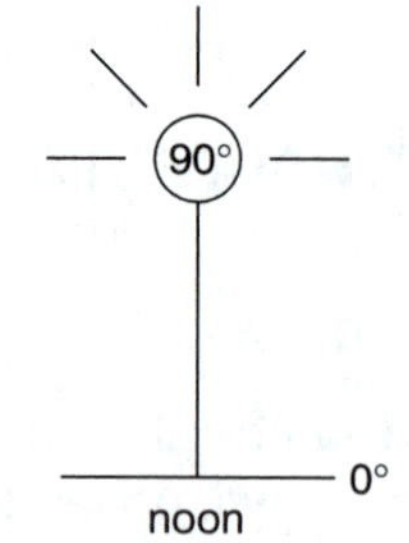

noon

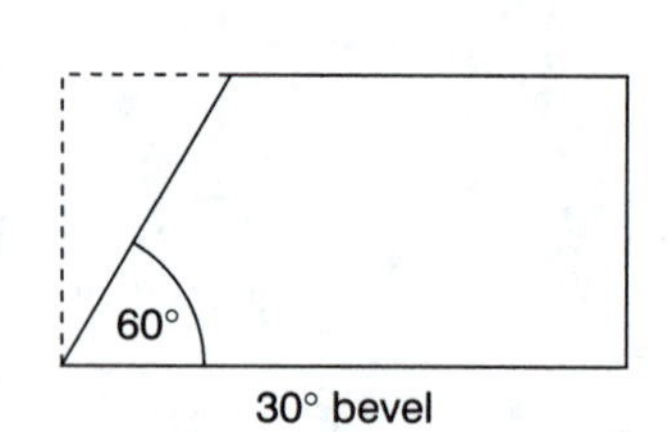

30° bevel

Find the angle (the number of degrees) that bisects the angle shown. Remember: *bisect* means to divide into two equal parts.

1. 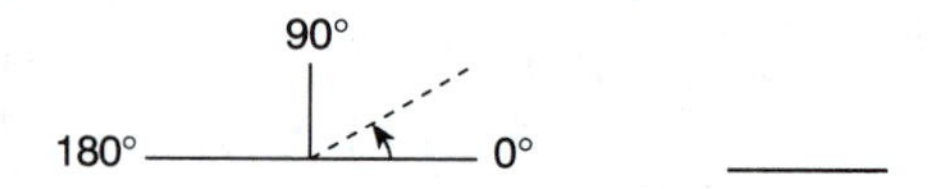

_____

2. 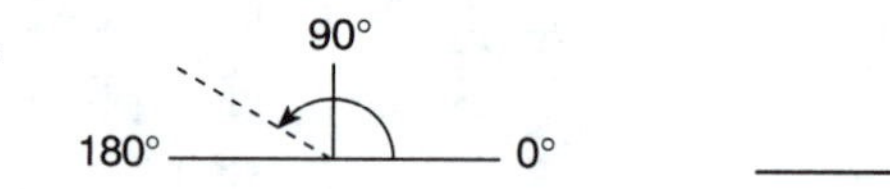

_____

3. 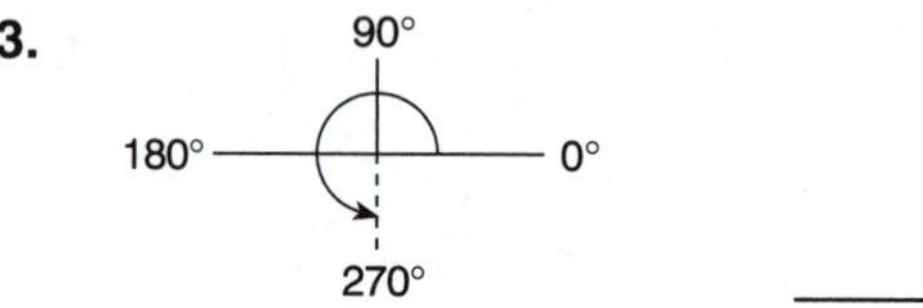

_____

4. 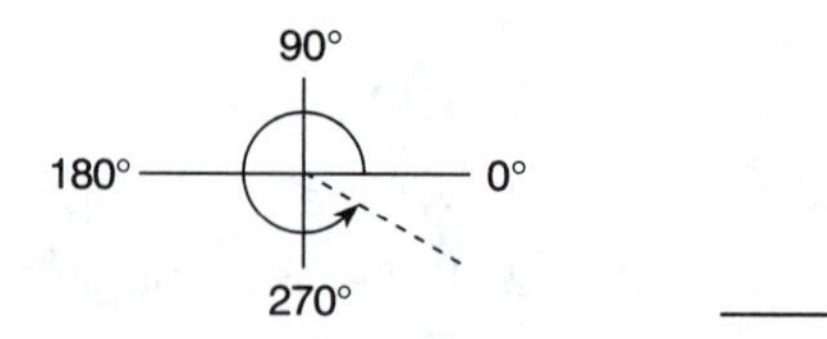

_____

Give the angle at which the electrode is positioned.

5. 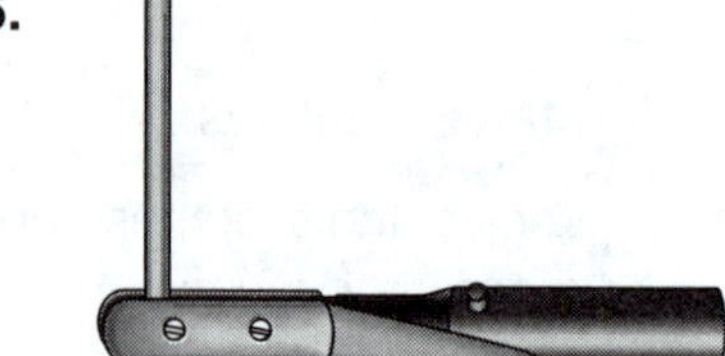

6.

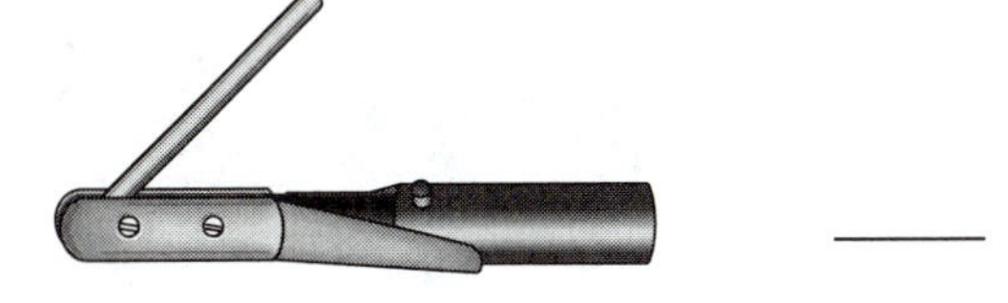

______

7.

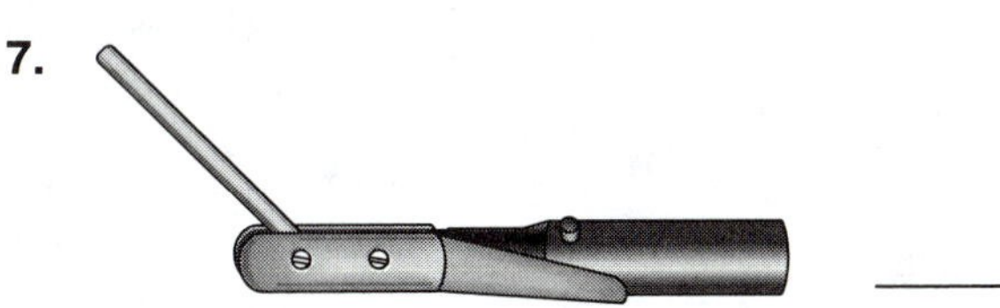

______

How many degrees of bevel are there in each of the following?

8. 75° ______

9. 60° ______

10. 30° ______

Find the groove angle for each of the following.

11. 60° 60° ______

12. 75° 75° ______

13. 67½° 67½° ______

14. 60° 60° ______

15. 45° 45° ______

Puzzlers

1. Turning the weldment over after the first pass, you can see that there is incomplete root penetration. What can you do to assure better penetration on the next root bead?
2. While vertical welding, the weld drips down. What can be done to eliminate the problem?
3. An injury occurs in the shop that is being attended to by your instructor. What should you do while waiting for the ambulance to arrive to pick up your classmate?

UNIT 5

Common Welding Problems

1. THE WELD BEAD

Two Basic Methods

The **weld bead** is deposited metal from a single pass, as shown in Figure 5–1. The stringer and the weave are the two basic methods for depositing filler metal. The **stringer** is a weld bead with little to no sideward movement of the electrode, usually kept within a width of two electrode diameters. By the use of the stringer the welder can maintain a uniform weld bead in a straight line to focus the heat for maximum root penetration. Laying stringers should be a basic exercise. Stringers allow the welder to concentrate on making weld beads in a straight line. Not all joints are welded using stringers; sometimes it becomes necessary to bridge a gap or do buildup using a sideward motion. Stringers are not always the choice when performing uphill welding in the vertical position.

The **weave** is a weld bead made from oscillation of the electrode from side to side in a patterned movement. The weave can be any one of several patterns (see Figure 5–2). The "C" pattern and the circular pattern (Figure 5–3) are two weaves that are easy

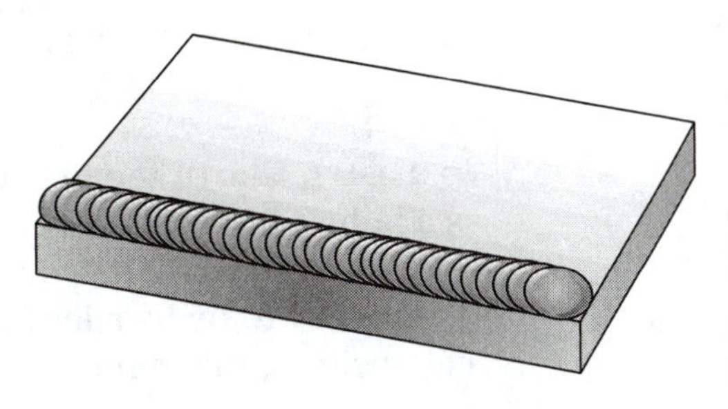

FIGURE 5–1 The stringer weld bead is positioned along the edge of the plate.

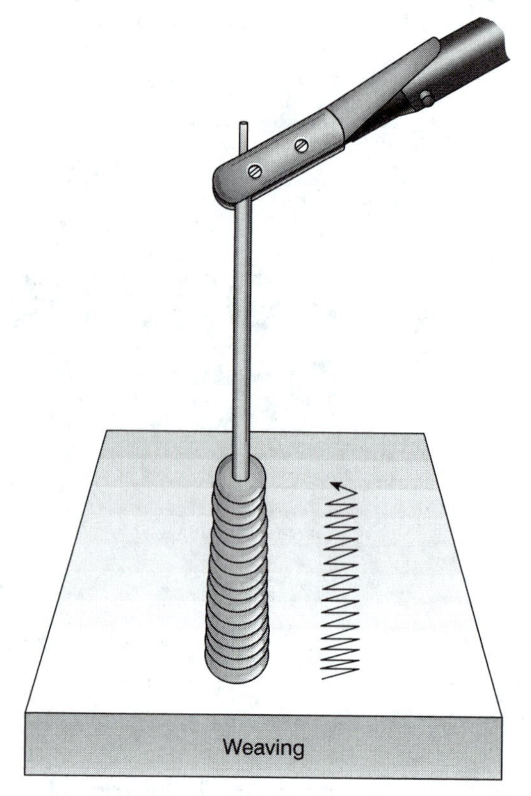

FIGURE 5–2 The weave technique can be done with several different patterns.

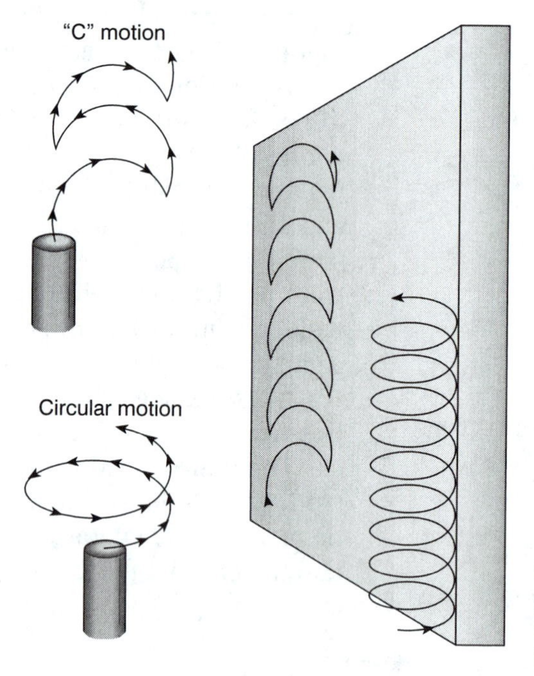

FIGURE 5–3 Two weave patterns are the "C" motion and the circular motion.

to master. One of the problems that can occur in making a weave is a failure to pause on the sides of the weld bead long enough to avoid **undercut.** When a welder fails to pause, the weld bead has a tendency to be heavier in the middle. When welding in the vertical position, this tendency will produce a sag, or "grapes," in the

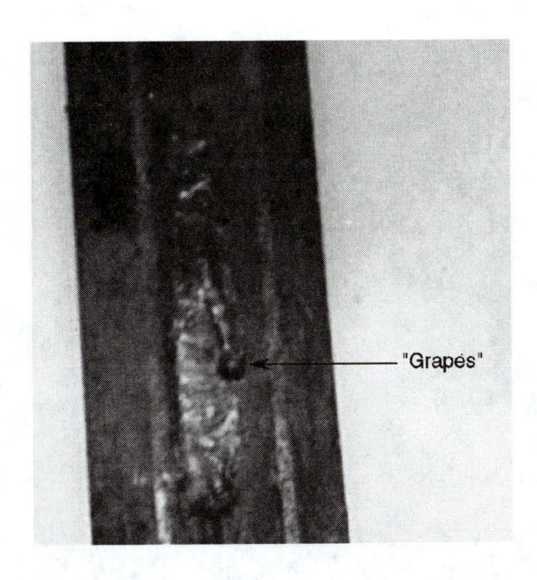

FIGURE 5–4 "Grapes" are a common problem in the vertical position when the welder loses control of the weld pool.

middle of the weld bead (see Figure 5–4). One remedy that assures the necessary pause is to count "one thousand one" on each side to give enough time to avoid undercut and make a smooth weld bead with even fill along the entire length of the weld. The weave tends to spread out the heat, which can restrict some of the root penetration into the base metal.

Welding demands control by keeping the hands steady. With practice a welder will develop some consistency in welding so that all the weld beads will begin to look similar. This will happen once the welder is able to achieve a consistent travel speed along the joint with every weld bead.

Failure of the Weldment

Unfortunately, the outcome from welding is not always what it should be. Every welder ought to be familiar with some common problems (Figure 5–5). Learn to recognize these problems to make the necessary corrections.

Several problems can cause the failure of a **weldment.** The weldment is an assembly of welded parts. Incomplete root penetration can result from too fast a travel speed, too low an amperage setting, or the improper preparation of the assembly for welding. In the act of welding the welder may not be aware that the root penetration is incomplete. Afterwards, the appearance of the weld bead can provide the evidence of incomplete root penetration. The appearance shows that the travel speed was too fast, or the amperage setting was too low.

If the travel speed is too fast, the weld bead has a high **weld face.** The weld face is the surface of the weld. With too fast a travel speed the weld bead is narrow with the weld pool ripple patterns coming to points (Figure 5–5, G). If the amperage setting is too low, the width of the weld bead is narrow. There is also a tendency to stick the electrode to the base metal, putting out the arc (Figure 5–5, B). Improper preparation of the assembly for welding

FIGURE 5–5 Comparison of acceptable weld bead with welding problems: (A) acceptable weld bead; (B) amperage setting too low; (C) amperage setting too high; (D) arc length too short; (E) arc length too long; (F) travel speed too slow; (G) travel speed too fast.

can include a failure to clean the base metal properly, or poor joint design (Figure 5–6).

Undercut is a groove left unfilled by deposited metal along the **weld toes** of the weld bead (Figure 5–7). The weld toes run the length of the weld bead, establishing the boundaries of the weld face. Undercut can be the result of an amperage setting that is too high (Figure 5–5, C), or too fast a travel speed (Figure 5–5, G). Undercut can also result when the arc length is too long (Figure 5–5, E), or when the electrode angle is improper. Undercut will occur if you do not pause at the end of the weld bead to fill in the **crater.** A crater is an undesirable depression left at the end of a weld bead. **Crater cracks,** or cracks that begin in the crater, will be the outcome (see Figure 5–8).

If the amperage setting is too high, it can cause the flux covering to break down, limiting its effectiveness. If the amperage setting is too high, the electrode cuts into the base metal, throwing metal away from the weld pool as spatter. The result is a wide bead with undercut along the toe on each side of the weld. Spatter takes the form of balls of metal forming on and around the weld bead. While not in itself affecting the strength of the weld, spatter

FIGURE 5-6 If practical, clean the metal before welding.

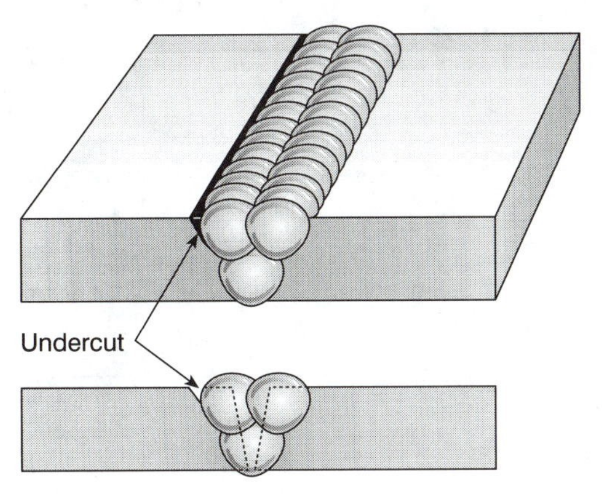

FIGURE 5–7 Undercut, a common welding problem.

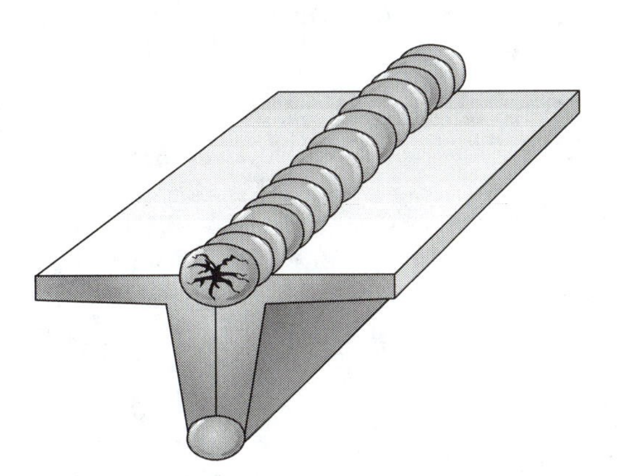

FIGURE 5–8 Crater cracks, another common welding problem.

gives a poor appearance and may be an indication of a more serious problem (Figure 5–5, C). Spatter can also result from maintaining an arc length that is too long (Figure 5–5, E), or poor quality (wet) electrodes.

Porosity (see Figure 5–9) refers to the pinholes (formed by trapped gas) that result from impurities in the base metal or in previous passes. Porosity can occur as a result of improper preparation or a failure to clean the weld (of moisture, rust, dirt) for the

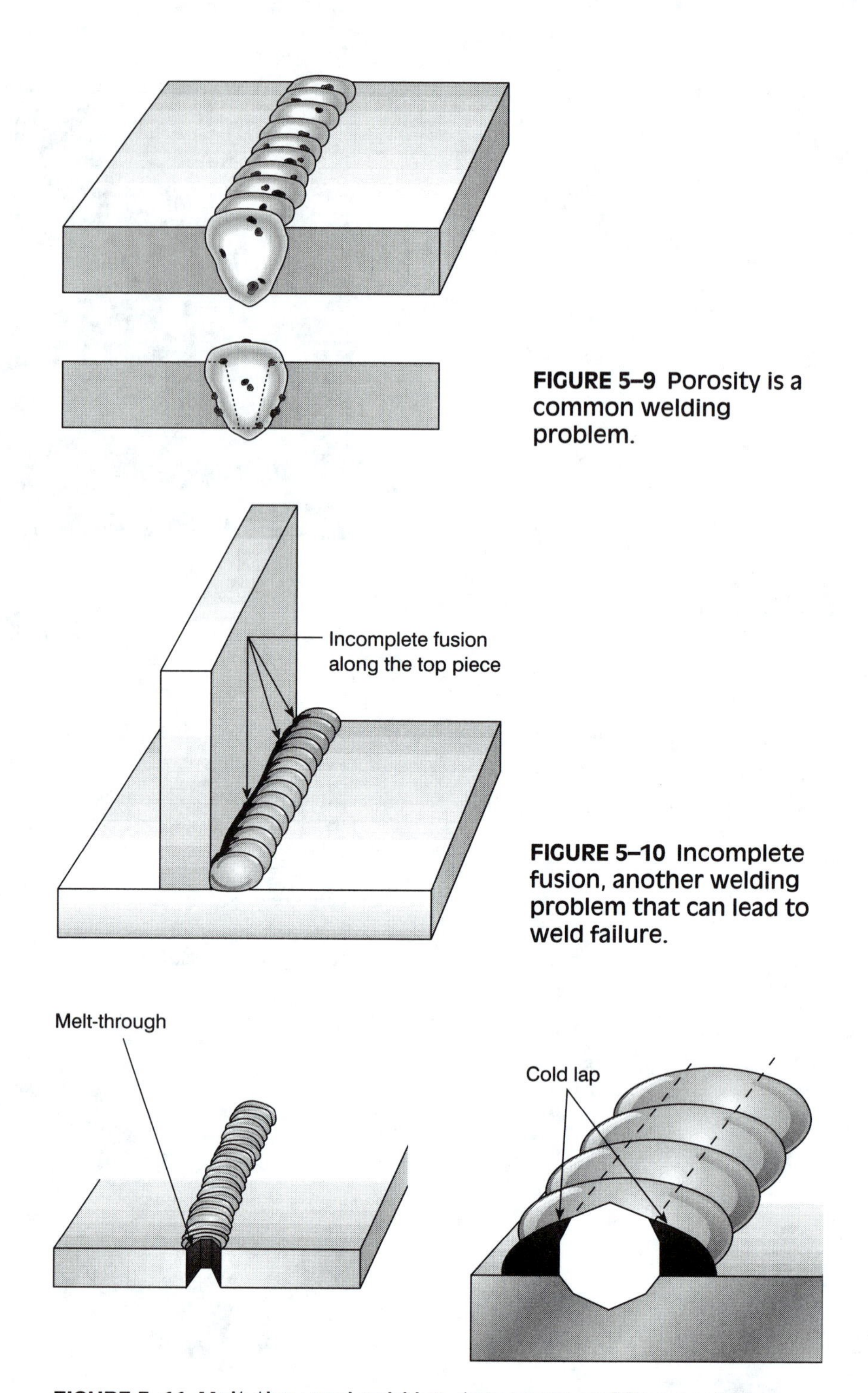

FIGURE 5–9 Porosity is a common welding problem.

FIGURE 5–10 Incomplete fusion, another welding problem that can lead to weld failure.

FIGURE 5–11 Melt-thru and cold lap, two more welding problems.

next pass. Porosity can be caused by too long or too short an arc length (Figure 5–5, D and E). Porosity can also happen when the amperage setting is too high (Figure 5–5, C), when there are drafty conditions, or when travel speed is too fast (Figure 5–5, G).

Additionally, **incomplete fusion** will occur if the travel speed is too fast. Incomplete fusion can also happen when the work angle is such that the weld bead ends up primarily on one member of the joint, shown in Figure 5–10.

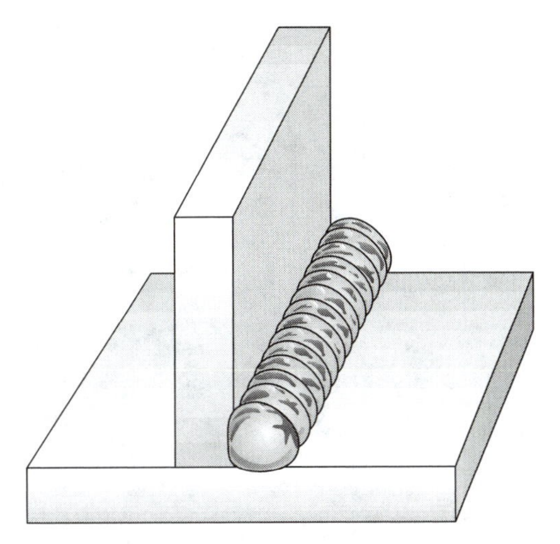

FIGURE 5–12 Slag inclusions are a common welding problem.

ARC FLASH

Grind off paint or rust before attaching the workpiece connection so the electric circuit can be completed.

Melt-thru or **cold lap,** both of which can cause incomplete fusion, results when the travel speed is too slow (Figure 5–5, F). These defects are illustrated in Figure 5–11.

Slag

Slag is nonmetallic material that can affect weld quality. Slag is the residue left when the covering on the electrode melts into the weld pool. When slag forms on the weld bead it usually can be removed easily upon cooling. The slag helps to protect the weld pool as it cools to keep contamination from the air being drawn into the weld as it solidifies.

Slag is also a nuisance, though, because it has to be removed by using a slag hammer and wire brush after each and every pass. If this time-consuming activity is not taken, the slag becomes trapped in the weld. This is called a **slag inclusion,** and it can affect the quality of the weld and possibly cause failure of the joint. Slag inclusions occur when nonmetallic material residue become trapped in the weld (Figure 5–12). Slag inclusions can result from a failure to remove the slag before making another pass, from incorrect amperage setting, or from improper welding techniques, including the wrong choice of electrodes or incorrect electrode angle.

Slag can become a problem if it begins to flow ahead of the arc, as in Figure 5–13. This can happen if the welder is not careful and use the incorrect electrode angle. This is more likely to happen when welding out of position. When welding downhill, the effects of gravity can cause the slag to flow down in front of the weld pool.

SAFETY REMINDER

Protect your sight by always wearing **safety glasses** under the welding helmet to keep hot slag from ending up in your eyes.

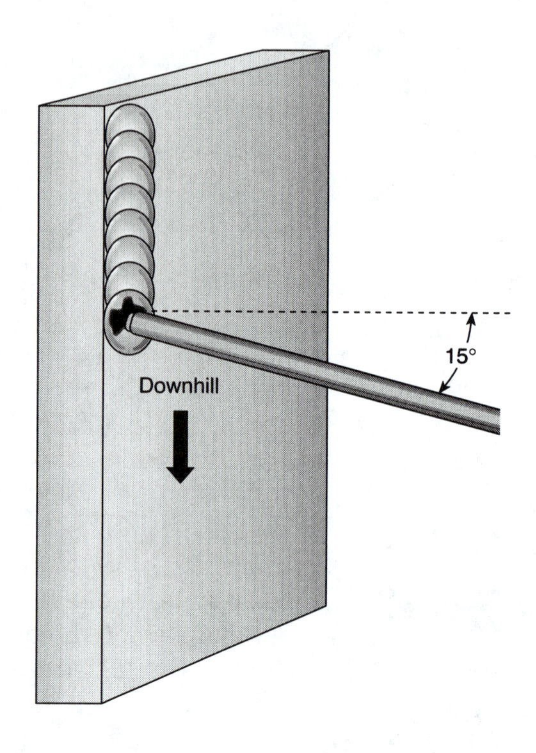

FIGURE 5–13 The downhill technique requires the welder to keep the weld pool from getting ahead of the electrode.

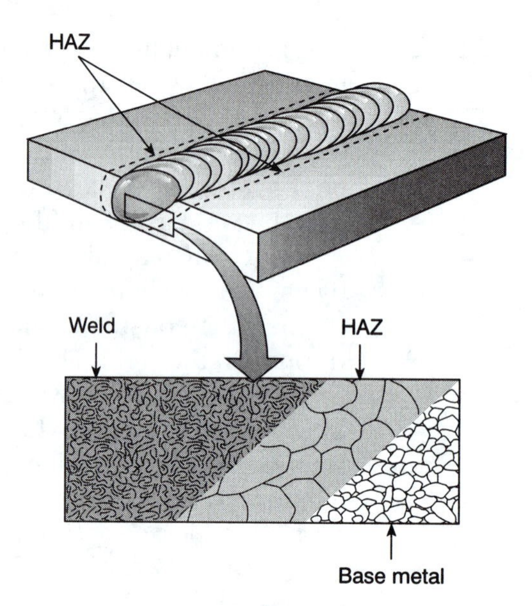

FIGURE 5–14 The steel is coarse grained in the heat-affected zone.

Cracking

Cracking of the weld can be a problem that is most likely to occur when welding medium-carbon steel and alloy steels. In some applications low-hydrogen electrodes and preheating the base metal to a temperature of approximately 300°–500° F (maintain 300° to 500° **interpass** temperature) can help to control the tendency of cracking in the **heat-affected zone (HAZ)** alongside the weld. There is a need to control the temperature during welding because weld metal subject to rapid cooling can change its inner structure and the structure of this HAZ area alongside the weld (see Figure 5–14).

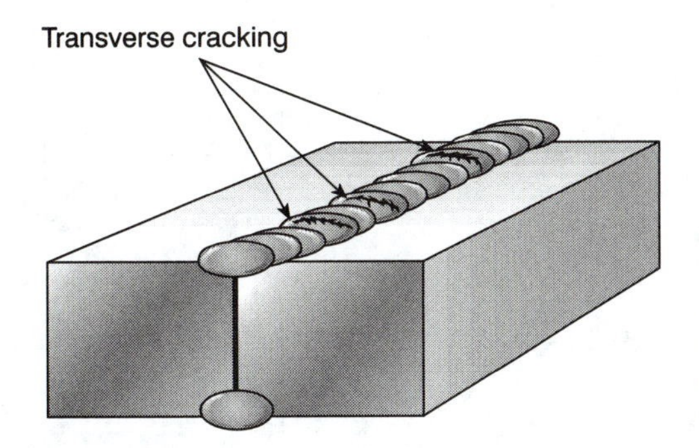

FIGURE 5–15 Transverse cracking is perpendicular to the length of the weld bead.

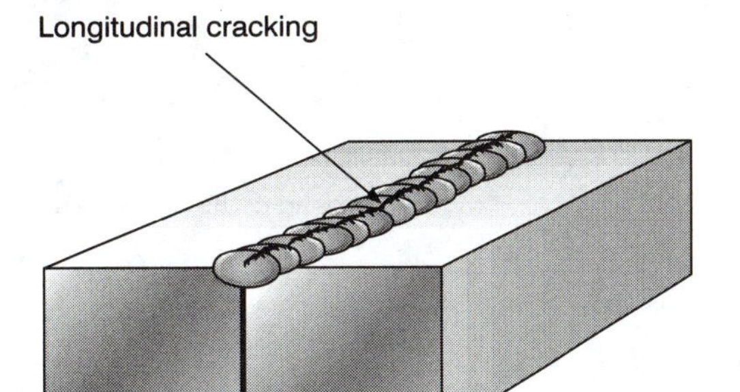

FIGURE 5–16 Longitudinal cracking is parallel to the length of the weld bead.

This HAZ area alongside the weld, even if never brought to melting temperature, is nonetheless affected by welding. The hardness of the weld will differ from the hardness of the HAZ area, which will differ from the unaffected base metal outside the HAZ area. A difference in hardness between these three areas is not a problem unless it exceeds specific limitations; then it can cause cracking. Other undesirable results occur when hardenable steels are hardened and hardened steels are softened in the HAZ area. **Underbead cracking** will also occur when gases such as carbon monoxide and hydrogen become trapped in the weld that cools before they can escape.

Preheating any steel over ½ inch in thickness sometimes works as a remedy to prevent rapid cooling that occurs in thick plate. Of course, it is always best to know the type of steel being welded to determine any special welding procedures that must be followed to prevent cracking. Furthermore, be sure to fill in each crater at the end of a weld to avoid crater cracks. Moving more slowly with a slight sideward motion to produce a larger weld bead will also reduce cracking, as will making multiple passes with stringers.

Transverse cracking is a defect that occurs in the weld from weld toe to weld toe (Figure 5–15). Transverse cracking can result from using the wrong electrode, making too small a weld for the size of the joint, or allowing the weld to cool too quickly. **Longitudinal cracking** is a defect that occurs in the weld face or the base metal parallel to the weld toes, as shown in Figure 5–16. Longitudinal cracking can happen when the joint is unable to move or when it moves too slowly as the weld cools. Choosing a

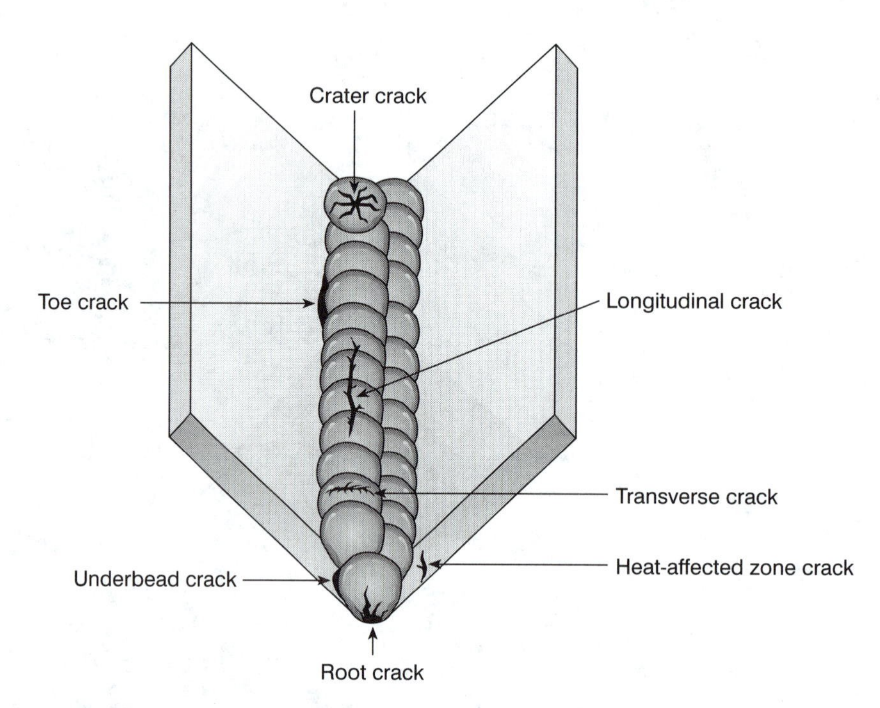

FIGURE 5–17 Several different kinds of cracks that can result from welding.

more ductile filler metal, selecting a less rigid joint design, or preheating and slow cooling of the joint will sometimes help.

Cracking results when metal under stress is relieved of energy. Welding steels that require special heating and cooling procedures are more susceptible to cracking than low-carbon steels. A welder should be aware of the different cracks brought about by welding (Figure 5–17).

Arc Blow

Arc blow is an erratic arc that refuses to go where the welder desires, causing spatter. Arc blow is formed by an unbalanced condition that creates a concentration of magnetic fields. When arc blow occurs, the welder may be unable to control where the weld bead is deposited (Figure 5–18). The bead does not flatten out, and the slag becomes difficult to remove, causing excessive spatter. While the arc may veer left or right during arc blow, normally the arc veers back and forth. The arc veers backward when welding toward the workpiece connection and forward when welding away from the workpiece connection.

Arc blow usually is not a problem with AC welding because the current changes direction constantly. Arc blow becomes more intense in DC as the amperage is increased. Arc blow can occur in deep groove welds, when welding into or out of corners, or when welding on magnetized metal.

The most common backward arc blow is recognized by excessive spatter and a narrow high bead with difficulty in removing the slag. Forward arc blow decreases in intensity as the distance increases from the workpiece connection. Forward arc blow has

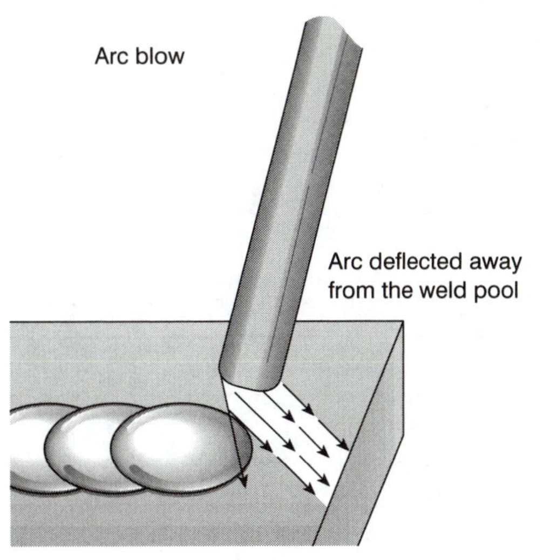

FIGURE 5–18 With an encounter of arc blow, the welder loses control of the arc.

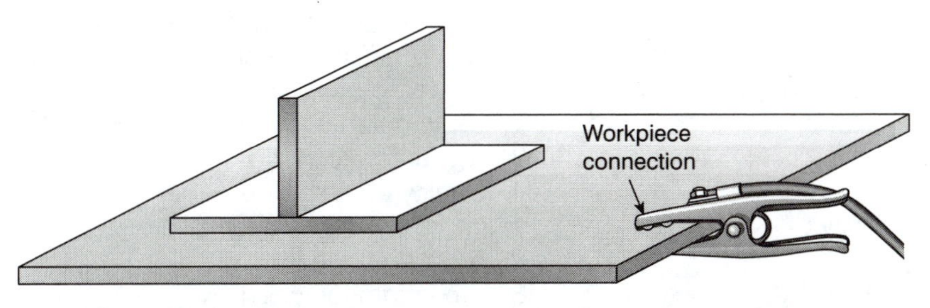

FIGURE 5–19 One way to limit arc blow is use of a large plate for attaching the workpiece connection.

a wide but uneven bead, with undercut and incomplete root penetration.

Here are some suggestions for dealing with arc blow. Arc blow can be controlled by switching to AC, especially above 250 amperes. Arc blow is sometimes controllable by changing the placement of the workpiece connection or by changing the welding direction (from left to right, or vice versa). Instead of placing the workpiece connection on a small project, try clamping the project to a large piece of metal on which the workpiece connection is attached, as shown in Figure 5–19. The welder might be able to control arc blow by reducing the arc distance and the travel angle while welding. Finally, the base metal may have to be demagnetized prior to welding.

REVIEW

1. What can cause porosity?
2. What can cause undercut?
3. What can cause spatter?
4. What can cause incomplete fusion?
5. What can cause melt-thru?
6. What can cause slag inclusions?
7. What are the effects of arc blow?

2. TROUBLESHOOTING GUIDE

A Summary of Shielded Metal Arc Welding Problems

TABLE 5–1 Troubleshooting guide for shielded metal arc welding.

Problem	*Cause*	*Remedy*
Incomplete root penetration	1. Travel speed too fast	Slow travel speed.
	2. Amperage too low	Increase the amperage.
	3. Improper preparation of the joint	Increase size of root opening.
Undercut	1. Amperage too high	Lower amperage.
	2. Travel speed too fast	Slow travel speed.
	3. Arc length too long	Shorten arc length.
	4. Incorrect electrode angle	Change work/travel angles.
	5. No pause to fill in crater	Pause and fill in crater.
Spatter	1. Amperage too high	Lower amperage.
	2. Undercut	*See undercut.*
	3. Arc length too long	Shorten arc length.
	4. Quality of electrode	Change electrode.
Porosity	1. Improper preparation of the joint	Clean before welding.
	2. Arc length too long	Shorten arc length.
	3. Arc length too short	Increase arc length.
	4. Amperage too high	Lower amperage.
	5. Drafty conditions	Eliminate drafts.
	6. Travel speed too fast	Slow travel speed.
	7. Quality of electrodes	Exchange electrodes.
Crater cracks	1. Undercut	Pause to fill in crater.
Incomplete fusion	1. Travel speed too fast	Slow travel speed.
	2. Incorrect work angle	Change work angle.
	3. Amperage too low	Increase the amperage.
Melt-thru	1. Travel speed too slow	Increase travel speed.
	2. Amperage too high	Lower amperage.
Cold lap	1. Travel speed too slow	Increase travel speed.
Cracked welds	1. Not enough weld	Increase the size of the weld.
	2. Weld cools too quickly	Slow rate of cooling.
	3. Electrodes do not match base metal	Match electrodes to base metal.
Slag inclusions	1. Failure to remove slag	Clean weld beads completely.
	2. Choice of electrodes	Change electrodes.
	3. Incorrect electrode angle	Change travel angle.
	4. Slag running in front of weld pool	Increase travel speed or change travel angle.
Sticking electrode	1. Amperage too low	Increase the amperage
	2. Quality of the electrode	Examine the electrode.
	3. Current choice (DCEP, DCEN, AC)	Change the electrode or change the current.
Arc blow	1. Unbalanced magnetic field	Change to AC. Reduce amperage. Shorten arc length. Change position of workpiece connection.

REVIEW

1. What are potential problems resulting from the following causes:
   a. Amperage is too high.
   b. Travel speed is too fast.
   c. Travel speed is too slow.
   d. Amperage is too low.
   e. Arc length is too long.
   f. Arc length is too short.
   g. Electrode angle is incorrect.
   h. Work angle is incorrect.

REVIEW QUESTIONS FOR UNIT 5

Multiple Choice

Choose the best answer.

1. Stringers:
   a. Are produced by using a sideward motion.
   b. Are used to bridge wide gaps.
   c. Are your only choice for uphill welding.
   d. None of the above.
   e. All of the above.
2. The weave method:
   a. Is not used for vertical position.
   b. Spreads out the heat.
   c. Is produced by a patterned movement of the electrode.
   d. a and b.
   e. a and c.
3. Practice:
   a. Allows you to develop consistency.
   b. Makes all your weld beads be similar.
   c. Is a waste of time.
   d. None of the above.
   e. a and b.
4. Undercut:
   a. Can occur when using a weave method.
   b. Is an unfilled groove along weld toes.
   c. Can result if amperage is too low.
   d. a, b, and c.
   e. a and b.
5. Spatter:
   a. Gives a poor appearance.
   b. Results if amperage setting is too low.
   c. Is metal from the electrode thrown away from the weld.
   d. Looks like little balls of metal.
   e. All of the above.
6. Melt-thru:
   a. Is a type of incomplete fusion.
   b. Results if travel speed is too slow.

c. Happens when amperage setting is too low.
d. None of the above.
e. All of the above.

7. Slag inclusion:
a. Is nonmetallic material residue trapped in the weld.
b. Is not the result of slag.
c. Is never a problem.
d. Can be avoided by welding downhill.
e. None of the above.

8. HAZ:
a. Cracking can be prevented by preheating in this area.
b. Is affected by rapid cooling.
c. Is part of the weld.
d. a and b.
e. a, b, and c.

9. Types of cracks include:
a. Toe crack.
b. Underbead crack.
c. Root crack.
d. Crater crack.
e. All of the above.

Short Answer

1. Give two causes of incomplete root penetration.
2. Describe a weld bead made with too fast a travel speed.
3. Write down four causes of undercut.
4. How can spatter be prevented?
5. What is porosity?
6. List some of the causes for incomplete fusion.
7. Give two causes of melt-thru.
8. What is slag?
9. What can cause slag to flow ahead of the arc?
10. What is arc blow?

ADDITIONAL ACTIVITIES

Solve a Mystery?

The detectives hovered above the steely victim. All that remained of the weldment was the fractured hulk of metal strewn about the site. The detective knew the pressure was on; if there wasn't an answer and soon, how many other victims would there be?

1. Clues: undercut along the weld toes of the vertical weld; not much spatter; sag in the middle of the weld. The solution is . . .
2. Clues: spatter all over the place; complete root penetration; undercut along the weld toes. The solution is . . .
3. Clues: ripple pattern comes to a point; narrow weld bead; high crown. The solution is . . .

4. Clues: transverse cracking along the weld; small weld bead. The solution must be . . .
5. Clues: spatter all over the place; root penetration, little to none; undercut along the weld toes. The solution has to be . . .
6. Clues: narrow weld bead; the electrode sticking to the base metal; incomplete root penetration. The solution you deduce is . . .
7. Clues: melt-thru; weld bead width too wide. Your solution to the crime is . . .
8. Clues: longitudinal cracking; rigid joint of thick plate. The mystery's solution has to be . . .

Writing Skills

1. Write a paragraph to explain the causes of arc blow and what can be done to minimize its effects.
2. Write a paragraph using the following words in sentences: *weld bead, stringers, weaving, vertical position, undercut, melt-thru, welder,* and *weld pool.*

Math

Read the tape:

1. ____________

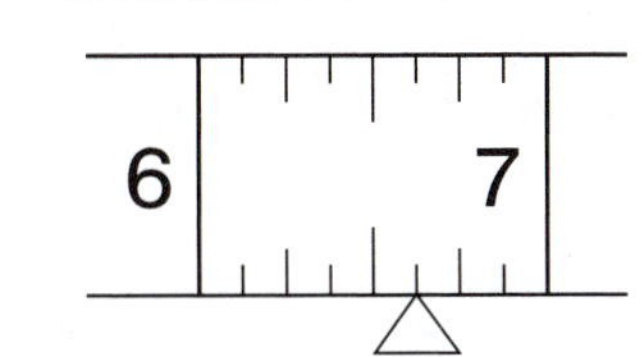

2. ____________

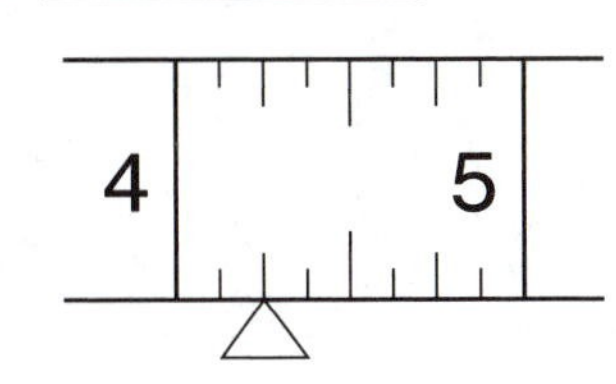

3. ____________

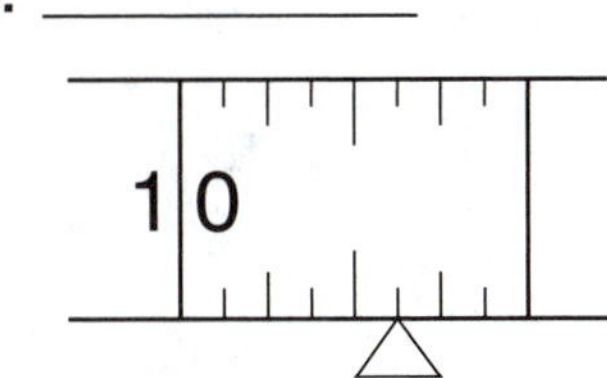

4. ____________

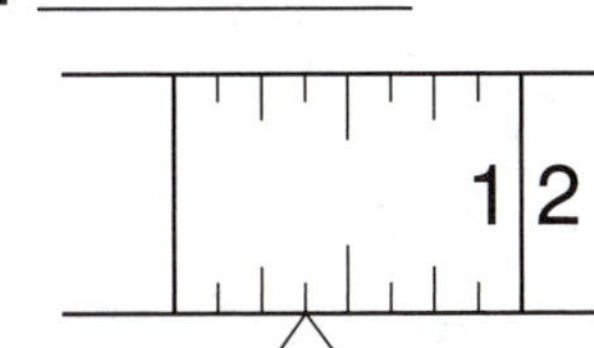

5. ____________

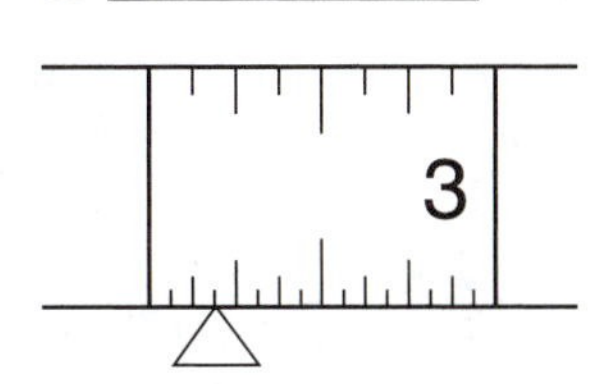

6. ____________

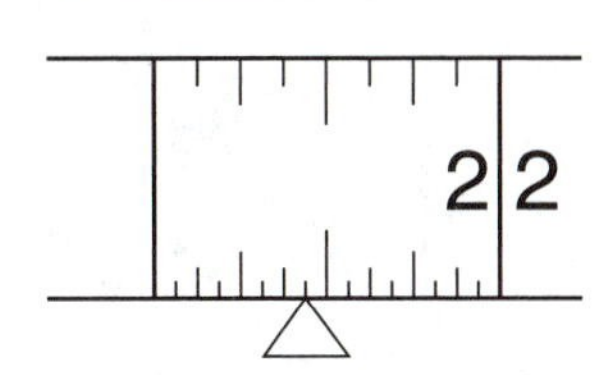

7. ____________

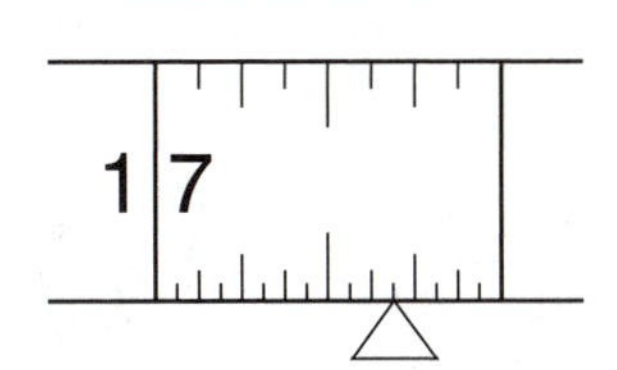

8. ____________

Puzzlers

1. Using DCEP to weld a tee joint in the horizontal position, you encounter trouble after a few inches of welding: the weld bead is laying only on the bottom plate. What can be done to solve this problem?
2. When would welding downhill in the vertical position be appropriate?
3. If the slag is difficult to remove, what are some of the possible causes?

UNIT 6

Welding Exercises

1. TWO POSITIONS FOR WELDING

Flat Position and Horizontal Position

The American Welding Society has laid out the boundaries for the four basic welding positions. This unit will focus on exercises for the **flat position** and the **horizontal position.** The appendix of this book contains exercises for the **vertical position** and the **overhead position.**

A position is considered flat if the plane of welding is from 0° to 15° (see Figure 6–1). Most welding is done in the flat position because working in that position allows more metal to be deposited into the joint in a single pass. The more metal put into a joint, the more work done in the shortest time at the least cost.

Welding in the flat position requires the least amount of skill. Yet the most time and energy learning shielded metal arc welding also involves the flat position. A beginning welder will have to invest some time to become proficient welding in the flat position.

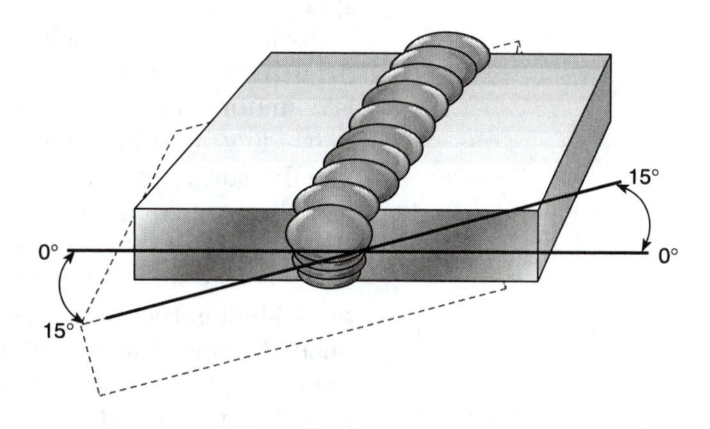

FIGURE 6–1 Plane for the flat position is from 0° to 15°.

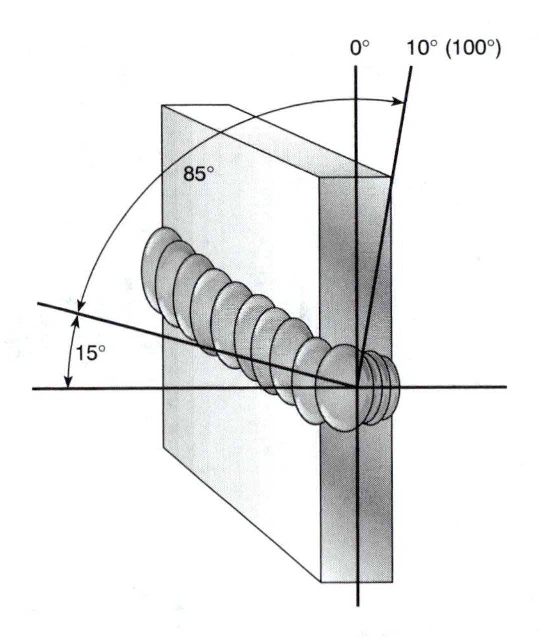

FIGURE 6–2 Plane for the horizontal position covers 80°.

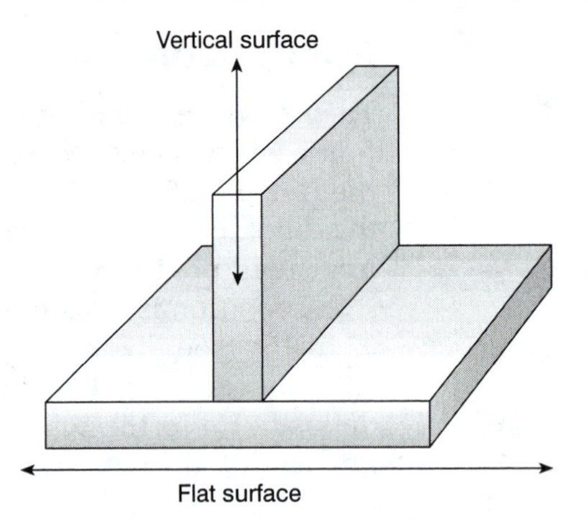

FIGURE 6–3 The tee joint.

However, the welder will benefit from the effort because the time spent learning the horizontal position will then be less.

A position is considered horizontal if the plane of welding is from 15° to 100°, as shown in Figure 6–2. The horizontal position is commonly used on projects. The **tee joint** is the best example of welding in the horizontal position. The weld is along a joint with one flat surface and one vertical surface (Figure 6–3).

All of the following exercises can be completed using ¼-inch and ⅜-inch thick steel. There are only two sizes necessary: plates ⅜" × 6" × 8" and plates ¼" × 3" × 8". For some exercises ⅜-inch thick steel is recommended, but all the exercises can be accomplished using ⅜-inch material. Six pieces of 6" × 8" plate and twelve pieces of 3" × 8" plate are enough to start working on the exercises presented in this unit.

REVIEW

1. What is the difference between the flat position and the horizontal position?
2. Sketch out a tee joint in three dimensions.

2. LAB EXERCISE NUMBER ONE

Laying a Bead

Each weld bead begins by placing an electrode into the electrode holder. By using a 135° angle, each electrode burns down to the stub end with very little waste. Be sure the workpiece connection is attached properly to establish an electric circuit. Set the power source for DCEP, if such capability is available.

Equipment and Material

Safety glasses	Wire brush	1 plate ⅜" steel, 6" × 8"
Welding gloves	Ear plugs	
Protective clothing	Grinder	Amperage: 90 to 110
Helmet	E6011 electrodes ⅛"	
Slag hammer		

Instructions

1. Clean the surface of the plate with a grinder.
2. If you are right-handed, move from left to right. If left-handed, move from right to left.
3. Tap the electrode on the plate, then raise it quickly. The electrode will stick if raised too slowly, and the arc will go out if raised too high. Maintain an arc length that approximately equals the diameter of the electrode. If the electrode sticks, release the electrode from the holder before raising your helmet.
4. Practice laying weld beads in a straight line. Start the first weld bead along the back edge of the plate, but in far enough to keep from burning off the edge.
5. Run stringers to keep the weld beads in a straight line. A stringer will not spread the heat. At this stage in your training, concentrate on being steady.
6. Continue to push the electrode into the weld pool to maintain the correct arc length while moving in the direction of travel.
7. Chip off the slag and brush the metal clean after each pass.
8. Test your work by visual inspection (see Figure 6–4). Continue to lay weld beads until each one is straight, ideally) ¼ inch wide, but no more than three wire diameters.
9. Check with your instructor for evaluation, if necessary, before moving to the next exercise.

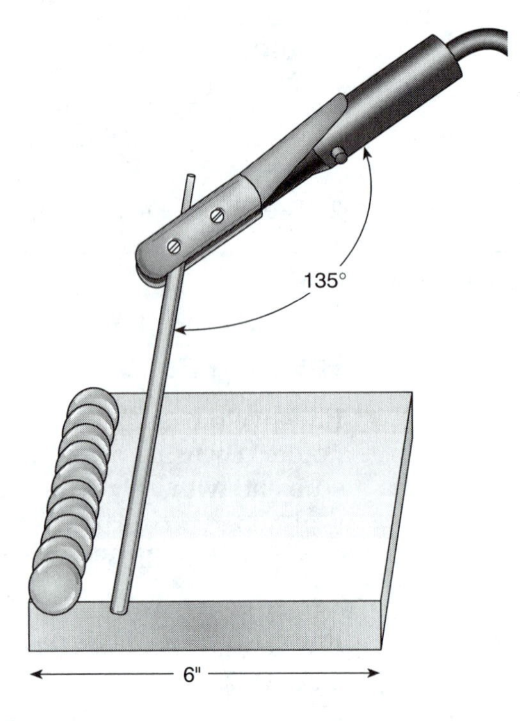

FIGURE 6–4 Position of the electrode using a 135° angle keeps the electrode holder out of the way and allows the welder to burn the electrode down to the stub end with little waste.

3. LAB EXERCISE NUMBER TWO

Padding Plate in Flat Position—Surfacing Weld

Vision is limited under the helmet. Lay the first weld bead along the back edge of the plate. Use the edge as a guide for keeping that first bead straight. If right-handed, begin on the left side of the plate. If left-handed, begin on the right side of the plate. Lay each weld bead so that it **laps** the previous weld bead by one-half its width, as in Figure 6–5.

If the plate has been cut to the dimensions recommended for this exercise, only a stub of electrode should be left after traveling the entire 8-inch length (see Figure 6–6). If some of the electrode remains, your travel speed was too fast. Upon reaching the edge at the end of the weld bead, break the arc momentarily, then quickly restart the arc, filling in the crater. This motion will prevent the edge of the plate from melting away. This exercise is an example of the kind of welding used to build up worn parts. There are electrodes designed for **buildup** to reduce wear.

Equipment and Material

Safety glasses	Wire brush	1 plate ⅜" steel, 6" × 8"
Welding gloves	Ear plugs	
Protective clothing	Grinder	Amperage: 90 to 110
Helmet	E6011 electrodes ⅛"	
Slag hammer		

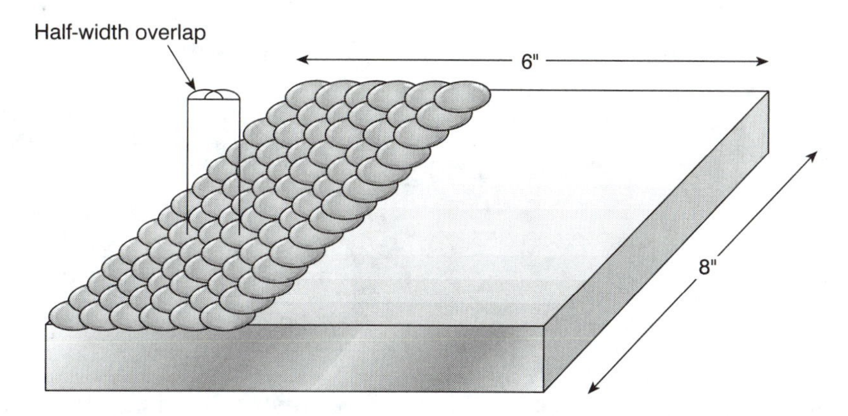

FIGURE 6–5 The padding plate dimensions with direction of travel.

FIGURE 6–6 Padding plate partially completed.

Directions

1. Clean the surface of the plate with the grinder.
2. Fill the plate completely with lapping weld beads. If the electrode is used up before reaching the edge of the plate, practice tying in the end of one weld bead with the beginning of the next weld bead, as shown in Figure 6–7. The challenge is to make each length of weld beads appear as one.
3. The ripples provide a pattern like fingerprints in each weld bead. The ripples should be rounded. Ripples that come to a point indicate a travel speed that is too fast. High, narrow weld beads suggest that the arc length is too short, DCEN is being used, or the amperage is too low.
4. If the arc length is too long, the weld bead will flatten out, the ripples will be uneven, and spatter will result. Spatter is the small drops or balls of metal from the electrode thrown away from the weld pool.
5. Look for straight weld beads with even ripples and consistent lapping.
6. Always tilt the top of the electrode 5° to 10° in the direction of travel. This helps prevent slag from being trapped in the weld bead.
7. Cool down the plate when the weld pool becomes wider than ¼ inch.

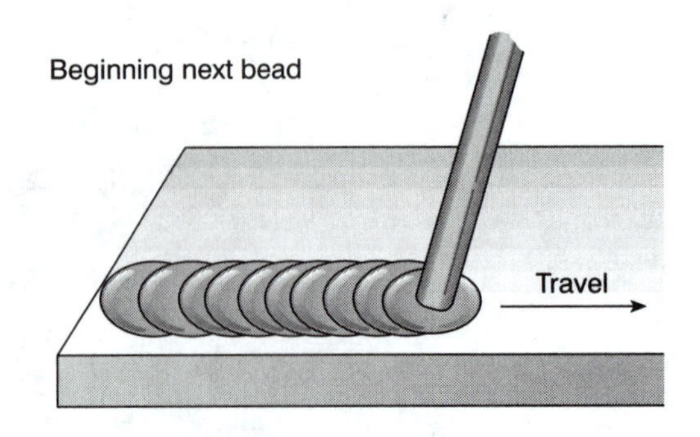

FIGURE 6–7 Tying one weld bead to another begins in the crater of the preceding weld bead.

8. Test is by visual inspection. Practice filling as many padding plates as needed to complete one with the appearance of quality welding.
9. Check with the instructor for evaluation, if necessary, before moving on to the next exercise.

4. LAB EXERCISE NUMBER THREE

Padding Plate in Flat Position—Surfacing Weld with Different Electrodes

This exercise requires a series of layers on a padding plate, as illustrated in Figure 6–8. Because each type of electrode melts differently, it is important to become proficient with some of the various types of electrodes. E6013 or E7018 are two suggestions of other electrodes you might try. E7018 requires a higher amperage setting, and its slag, much like that from E6013, chips off easily.

Remember that the proper travel speed, amperage setting, arc length, and electrode angles (work angle and travel angle) are important for welding technique.

Equipment and Material

Safety glasses	Wire brush	1 plate ⅜" steel, 6" × 8"
Welding gloves	Ear plugs	Amperage: *see Table 4–1*
Protective clothing	Grinder	
Helmet	Various electrodes\sizes	
Slag hammer		

Directions

1. Clean the surface of the plate with a grinder.
2. Use the slag hammer and wire brush on every weld bead before laying a lapping bead.
3. Fill all craters at the edge of the plate.
4. Keep a tight or close arc length to reduce spatter.
5. Run one layer of lapping weld beads using one type of electrode. Look for straight weld beads with even ripples.

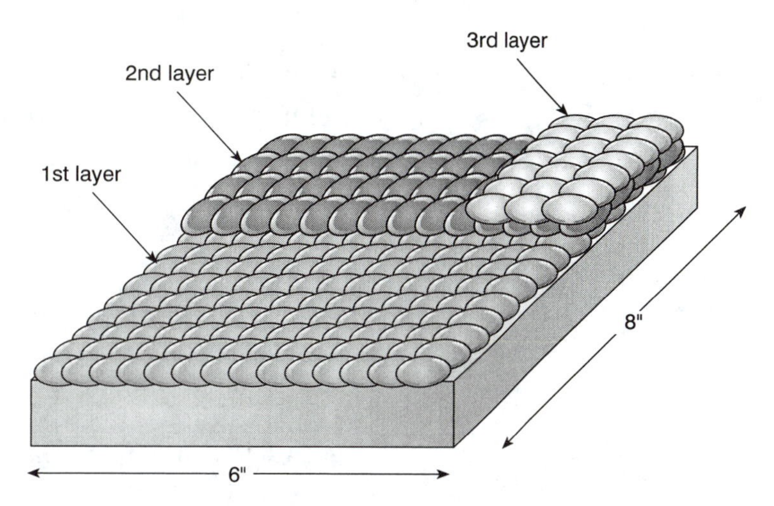

FIGURE 6–8 Surfacing weld with different electrodes; note changing the direction of travel with each layer.

6. Practice until you have a quality padding plate.
7. Complete a second layer using a different electrode or different size electrode. Did you have to adjust the amperage setting?
8. Complete a third layer using another electrode. Must you adjust the amperage setting with this change of electrode?
9. Test is by visual inspection. Practice this exercise until you are confident the appearance of your padding plate is that of quality welding.
10. Check with the instructor for evaluation, if necessary, before moving on to the next exercise.

5. LAB EXERCISE NUMBER FOUR

Corner Joint in Flat Position—Fillet Weld

Use ¼-inch angle or weld two pieces of ¼-inch plate together in the dimensions recommended. This exercise provides an opportunity to practice the kind of welding required in fusing parts together using the fillet weld (Figure 6–9). As in the padding plate exercises, lay stringer weld beads so that they lap one another in a straight line. No space should be left between the weld beads where slag can be trapped.

The **tack weld** is introduced with this exercise, if two pieces of steel are to be joined together. Tack welds are short temporary weld beads used to hold a weldment in position to complete welding. Yet tack welds are very important in any fabrication project. All the parts that are essential to the weldment must be tack welded first.

Equipment and Material

Safety glasses
Welding gloves
Protective clothing
Helmet
Slag hammer
Wire brush
Ear plugs
Grinder
E6011 electrodes ⅛"
One 1" × 1" × ¼" angle × 8" or two pieces of ¼" plate by 8"
Amperage: 90 to 110

Directions

1. Clean the surface with the grinder.
2. Tack weld the angle (or two pieces of plate) to a scrap padding plate.
3. Lay the weld beads in each layer of the fillet weld by moving from one side to the other, lapping each weld bead by one-third to one-half.
4. Build up layer upon layer.
5. Remove all slag from each weld bead before laying another weld bead.
6. Make each layer as smooth and clean as the buildup in the padding plate exercises.
7. Complete three to five layers with the ⅛" E6011 electrode.
8. Complete an additional two layers with another type of electrode.
9. If slag inclusions are visible, do a better job of cleaning each weld bead before laying another weld bead.
10. Test is by visual inspection. Practice this exercise until confident that the appearance of your fillet weld is of quality.
11. Check with the instructor for evaluation, if necessary, before moving on to the next exercise.

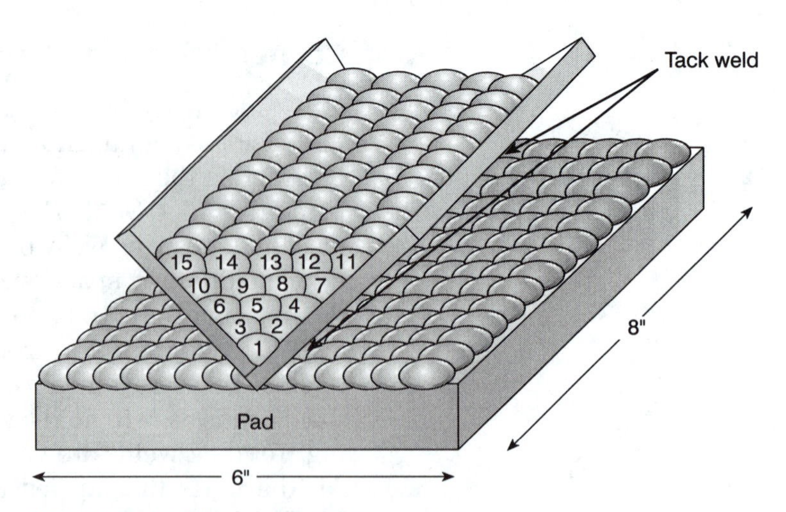

FIGURE 6–9 Fillet weld in flat position.

6. LAB EXERCISE NUMBER FIVE

Outside Corner Joint in Flat Position—Mechanical Testing: Groove Weld

The outside **corner joint** is a popular joint design in many projects. The **groove weld** is used for joining metal in which the edges forming the joint are welded together. There should not be a problem determining the size of the weld; it is usually equal to the thickness of the metal in the joint.

This exercise provides an opportunity to test the strength of your welding. Judging the appearance of the weld is important. Performing a **destructive test** on the joint gives even more of an indication as to how well you are doing in these exercises.

Equipment and Material

Safety glasses	Ear plugs	Amperage: 90 to 110
Welding gloves	Grinder	
Protective clothing	E6011 electrode, ⅛"	
Helmet	Two pieces of ⅜" plate steel, 3" × 8"	
Slag hammer		
Wire brush		

Directions

1. Clean the surfaces, including the edges that will be welded together.
2. Tack weld the two pieces together as shown in Figure 6–10. Be sure to remove all slag.
3. Make stringer weld beads. The first weld bead should run right over the tack welds.

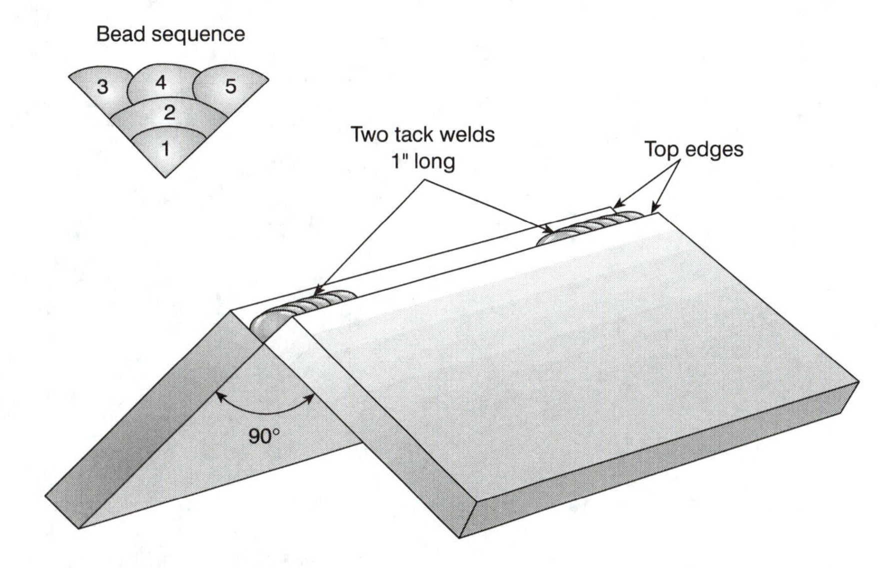

FIGURE 6–10 Outside corner joint in flat position.

4. Note the weld bead sequence. Four weld beads are required to complete the weld.
5. On the **cover pass** (final layer of weld beads), remove the top edges of both pieces.
6. Cool. Then flatten, if possible. A **press brake** will do the job.
7. Further test is by a visual inspection. Look for root penetration into the edges of the joint without cracking.
8. If your weld held up under the stress of testing without breaking, you are ready to move on.
9. Check with the instructor for further evaluation, if necessary, before moving on to the next exercise.

7. LAB EXERCISE NUMBER SIX

Padding Plate in Horizontal Position—Surfacing Weld

With an introduction of the horizontal position in this exercise, the work angle and travel angle of the electrode become important (see Figure 6–11). When starting out to weld in the horizontal position, there is a tendency to move too quickly across the plate for fear that the weld pool will drip, as shown in Figure 6–12. The welder should not be overly concerned, and should fight this feeling. The proper position of the electrode helps to counteract gravity. Move steadily, but not quickly, across the plate.

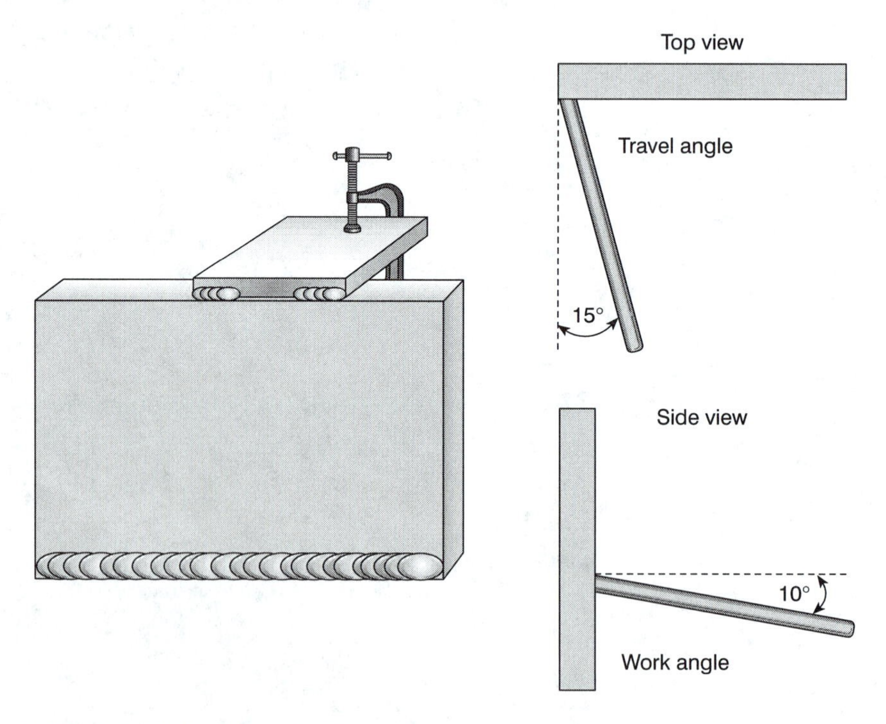

FIGURE 6–11 Padding plate positioned for welding; note travel angle and work angle.

To hold the plate vertically for welding, tack weld a piece of steel on the back side. This piece of steel can be used to hold the plate in a fixture. Otherwise, tack the plate to a metal table or weld the plate to another length of steel that is lying on a table. Try to provide some height. That way the plate can be closer to eye level, a more comfortable position for welding.

Equipment and Material

Safety glasses	Wire brush	One plate ⅜" steel, 6" × 8"
Welding gloves	Ear plugs	
Protective clothing	Grinder	Amperage: 90 to 110
Helmet	E6011 electrode, ⅛"	
Slag hammer		

Directions

1. Beginning at the bottom, lay lapping weld beads until the plate is full (as in a padding plate). Make one layer with the E6011. Later you can try another type of electrode or another size electrode for the second layer.
2. Use the previous weld bead as a shelf to place each following weld bead.
3. Remember that the plate might have to be cooled occasionally between passes, or the amperage might have to be lowered on the power source. Otherwise, the plate will become too hot, causing the weld pool to drip.
4. To begin, tilt the electrode holder 10° to 15° in the direction of travel. Adjust the work angle of the electrode by moving the electrode holder downward 10° to counteract gravity (see again Figure 6–11).
5. The ripples in each weld bead should be rounded. Ripples that come to a point indicate a travel speed that is too fast. High, narrow weld beads indicate an arc length that is too short, DCEN (wrong current choice), or amperage too low.

FIGURE 6–12 Surfacing weld in horizontal position.

6. Look for straight weld beads with even ripples and sufficient lapping.
7. Test by a visual inspection. Practice as many padding plates as necessary to obtain quality welding.
8. Check with the instructor for evaluation, if necessary, before moving on to the next exercise.

8. LAB EXERCISE NUMBER SEVEN

Tee Joint in Horizontal Position—Fillet Weld

The tee joint is very common in welding projects. In this challenging exercise, two pitfalls to avoid are piling up weld metal and poor lapping. Piling up filler metal on the bottom piece can result in joint failure. A cross-section of such a weld would show the solution to this problem when welding usually requires an even distribution of filler metal between both members of the weldment. The first weld bead of each layer is critical. Two-thirds of this weld bead should cover the bottom weld bead of the previous layer. Poor lapping within the completed weld results in valleys and ridges. Sufficient lapping results in a smooth surface, as in Figure 6–13.

The work angle is important in this exercise, as pictured. Each weld bead should be tight against the other weld beads of the joint. Watch out for arc blow, making any adjustments as needed to avoid it.

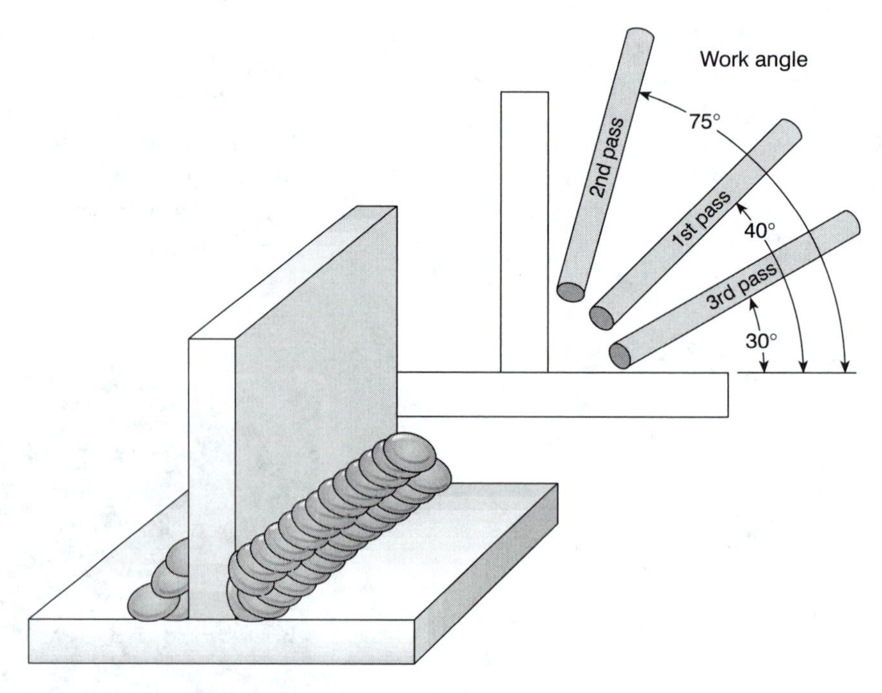

FIGURE 6–13 Fillet weld in horizontal position.

Equipment and Material

Safety glasses	Wire brush	Two pieces of plate ¼" to ⅜" steel, 3" × 8"
Welding gloves	Ear plugs	Amperage: 90 to 110
Protective clothing	Grinder	
Helmet	E6011 electrode, ⅛"	
Slag hammer		

Directions

1. Begin each pass (layer) from the bottom. Lay 10 to 15 weld beads on each side, using 6011 on one and some other electrode on the other side. Alternate from side to side. Cool to reduce the effects of heat.
2. Remove slag completely before laying the next weld bead.
3. Test is by visual inspection. Look for a smooth fillet weld with straight weld beads. Examining from the side should show an even distribution of fill between the top and bottom plates.
4. Practice this exercise until you are making quality welds.

Testing

Inspection begins with a visual examination of the weld. A visual inspection is made every time you lay a weld bead. Learn from your mistakes by observations made both during and after the completion of welding. As you are welding, watch the weld pool, making immediate adjustments to what you see. The visual inspection is the first and most inexpensive way of examining the weld.

Obviously, the visual inspection is not always enough. There are both destructive and more sophisticated **nondestructive** methods of examining welds. Destructive testing can be costly because the welded joint is destroyed in the testing process. The American Welding Society (AWS) has developed procedures for testing a tee joint. Run a destructive test on a tee joint by following the instructions listed here; then complete the AWS visual test:

1. Complete one side of a tee joint with three weld beads; ¼-inch leg size.
2. First, make a visual inspection, examine for excessive undercut, porosity, and weld bead lapping.
3. Put in a vice and hammer the vertical plate over (press brake preferred), Figure 6–14.
4. According to the American Welding Society's *Structural Welding Code—Steel D1.1*, a visual test of the fractured fillet weld should show no single **discontinuity** (such as porosity or slag inclusion) greater than 3/32 inch, with a sum of all discontinuities no greater than ⅜ inch for 6 inches of weld; 1 inch on each end is discarded.
5. Check with the instructor for evaluation, if necessary.

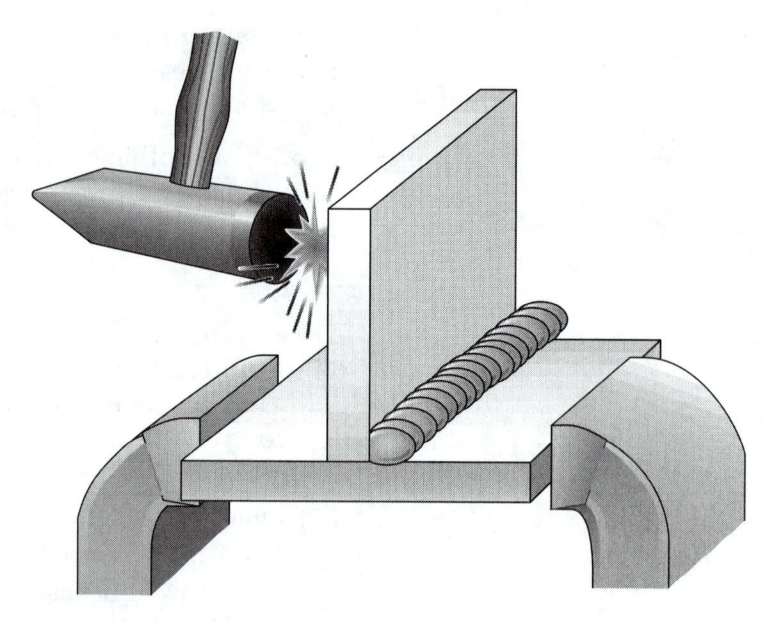

FIGURE 6–14 Apply force of hammer on side opposite the weld is one method for destructive testing of the weld.

REVIEW QUESTIONS FOR UNIT 6

Multiple Choice

Choose the best answer.

1. The flat position in welding:
   a. Takes the most time to learn.
   b. Requires the least skill.
   c. Allows more metal to be deposited.
   d. Takes less time than the overhead position.
   e. All of the above.
2. Necessary equipment for welding includes:
   a. Safety glasses.
   b. Slag hammer.
   c. Wire brush.
   d. Welding gloves.
   e. All of the above.
3. A surfacing weld:
   a. Is another name for a butt joint.
   b. Builds up the surface of the plate.
   c. Is not a padding plate.
   d. All of the above.
   e. None of the above.
4. Ripples:
   a. Are the patterns of each weld bead.
   b. Are a type of slag.
   c. If pointed, indicate that the travel speed was too fast.
   d. a and c.
   e. None of the above.
5. The electrode angle:
   a. Is the same as the work angle.
   b. Should always be 90°.

c. Is the same as the travel angle.
d. a and c.
e. None of the above.

6. Tack welds:
a. Are important in fabrication.
b. Are permanent weld beads.
c. Hold the weldment in position.
d. a and b.
e. All of the above.

7. Visual inspection:
a. Requires special technology.
b. Takes too much time.
c. Is the first test applied to any weld.
d. Is sometimes used.
e. None of the above.

8. The Destructive test:
a. Is the least expensive way to test a weld.
b. Is another type of nondestructive test.
c. Destroys the welded joint.
d. Is never used.
e. Is not recognized by the AWS.

9. Discontinuities include:
a. Porosity.
b. Slag inclusion.
c. Weld bead.
d. a and b.
e. None of the above.

Short Answer

1. What is the American Welding Society?
2. Why is most welding completed in the flat position?
3. Give one major reason for including a grinder in these exercises.
4. What type of inspection is given to every weld?
5. When is a surfacing weld used?
6. Name one problem that is prevented by tilting the electrode slightly in the direction of travel.
7. Name three things that are important for welding technique.
8. What is the function of a tack weld?
9. What can happen if all the slag is not removed before making another pass?
10. How is the destructive test different from a nondestructive test?

ADDITIONAL ACTIVITIES

Riddles

1. From my "position" I scare folks who are unnecessarily worried about things falling on their heads.
2. I am an unfilled groove along the weld toes.

3. Smart welders use my approach on thin metal when melt-thru can be a problem.
4. Some, not mentioning any names, consider me a nuisance, but then they probably aren't listening or hearing any better either.
5. I like to knock slag around.
6. I am used from side to side, but what the heck, I don't mind.
7. People say I destroy things, but not so fast—I'm only performing a service of testing welds.
8. I am a credible standard of excellence.
9. I start the welding process rolling, but I should be kept in my place, the V-groove or welding area, that is.
10. Speed up, slow down. Welders have to keep me in mind during welding.
11. I prevent distortion during welding exercises that will be tested.
12. Without me your eyes aren't going to be worth very much.

Writing Skills

1. Write a paragraph describing the components of the helmet used for welding.
2. Explain how the work angle is different from the travel angle.

Math

A welder should know some of the fractions common in measuring. Normally a welder should be able to fabricate things that are within a tolerance of ⅛ inch. Yet in taking measurement with a tape (cutting and fitting together for tack welding), the welder should be able to measure within ¹⁄₁₆ inch.

Reduce the following two fractions to sixteenths.

1. $\frac{4}{32} = \frac{\quad}{16}$ **2.** $\frac{6}{64} = \frac{\quad}{16}$

Reduce the following two fractions to eighths.

3. $\frac{12}{32} = \frac{\quad}{8}$ **4.** $\frac{16}{64} = \frac{\quad}{8}$

Answer the following.

5. $\frac{2}{16} = \frac{\quad}{8}$ **6.** $\frac{6}{16} = \frac{\quad}{8}$ **7.** $\frac{10}{16} = \frac{\quad}{8}$

8. $\frac{7}{8} = \frac{\quad}{16}$ **9.** $\frac{3}{8} = \frac{\quad}{16}$ **10.** $\frac{7}{16} = \frac{\quad}{8}$

11. $\frac{1}{8} = \frac{\quad}{16}$

A welder should know the decimal equivalents for eighths and sixteenths.

Example: $\frac{1}{16}$ = 0.0625 (1 ÷ 16 = 0.0625)

Example: $\frac{1}{8}$ = 0.125 (1÷ 8 = 0.125)

12. $\frac{3}{16}$ = **13.** $\frac{5}{16}$ = **14.** $\frac{5}{16}$ = **15.** $\frac{7}{16}$ = **16.** $\frac{9}{16}$ =

17. $\frac{11}{16}$ = **18.** $\frac{13}{16}$ = **19.** $\frac{3}{8}$ = **20.** $\frac{5}{8}$ = **21.** $\frac{6}{8}$ =

22. $\frac{7}{8}$ =

UNIT 7

Joint Design

1. FIVE BASIC JOINTS

Choice of Joint Design

Welding begins with the simple idea of joining two pieces of metal together in a joint. There are five basic joints used in welding. Whenever two pieces of metal are joined, they fit into one of these basic joints. The five joints at the foundation of every welding project are the **butt joint,** the **tee joint,** the **lap joint,** the **corner joint,** and the **edge joint** (see Figure 7–1).

Selection of the joint design is important in the fabrication of any project. The idea is to achieve maximum strength in the joint with efficiency of the weldment. That is, joints are designed to carry out given tasks while requiring the least amount of welding. For example, fabrication of a container to hold a liquid would find the corner joint a more practical choice than a tee joint. A corner joint assures a seal, saves material, and looks attractive (Figure 7–2).

Exercises in Unit 6 emphasize the corner joint and the tee joint. These two joints plus the lap joint cover the three joints most easily welded. There should never be a mystery as to how much weld is enough. Size your welds equal to the thickness of the thinnest metal in the joint.

Fillet Weld

To become proficient at welding, the welder should know something about the anatomy of the welds that are being made. The **fillet weld** is a commonly used weld, as shown in a tee joint in Figure 7–3. The fillet weld is a weld that joins the surfaces of two pieces, usually at right angles to each other. The fillet weld is

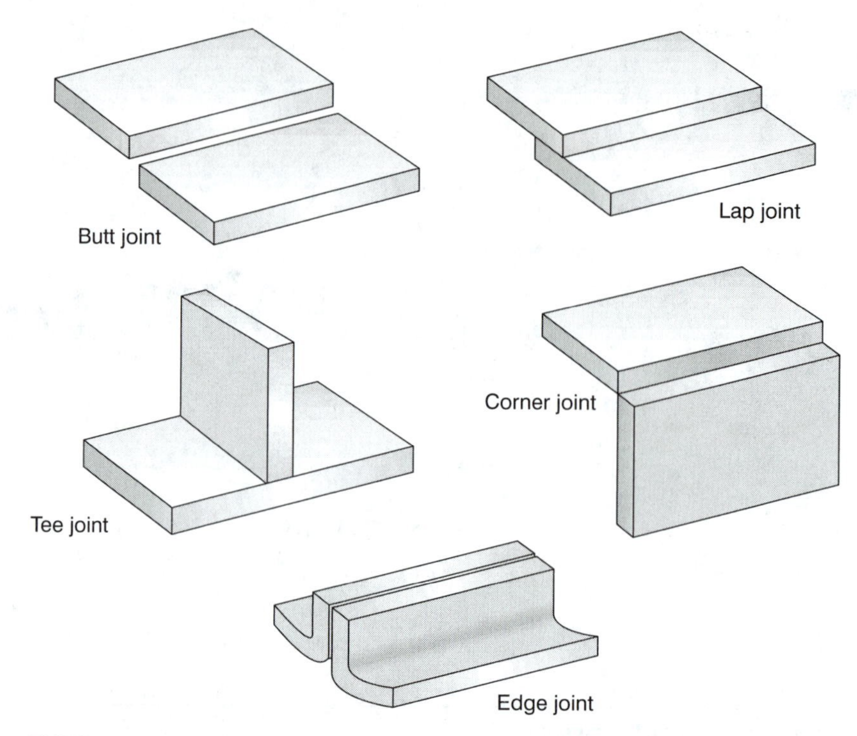

FIGURE 7–1 The five basic joints used in welding.

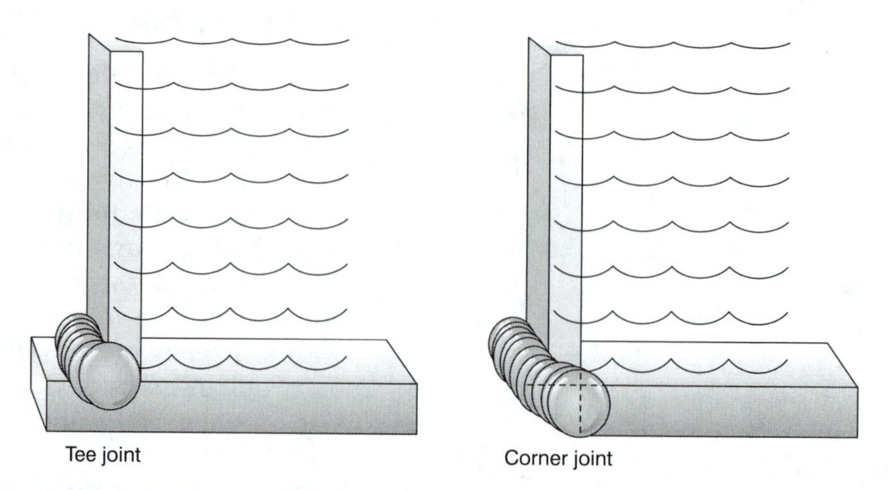

FIGURE 7–2 Use of a corner joint for sealed container applications is generally preferred over the tee joint.

popular because it does not require the added preparation necessary for the groove welds. Often the only preparation of the base metal for a fillet weld is grinding away surface materials that could interfere with the quality of the weld.

Groove Welds

Used most often in the butt joint, the **groove weld** requires the most welding skill. The groove weld demands added preparation with the aim being to achieve complete root penetration. This means the weld can be viewed as penetrating completely (100%)

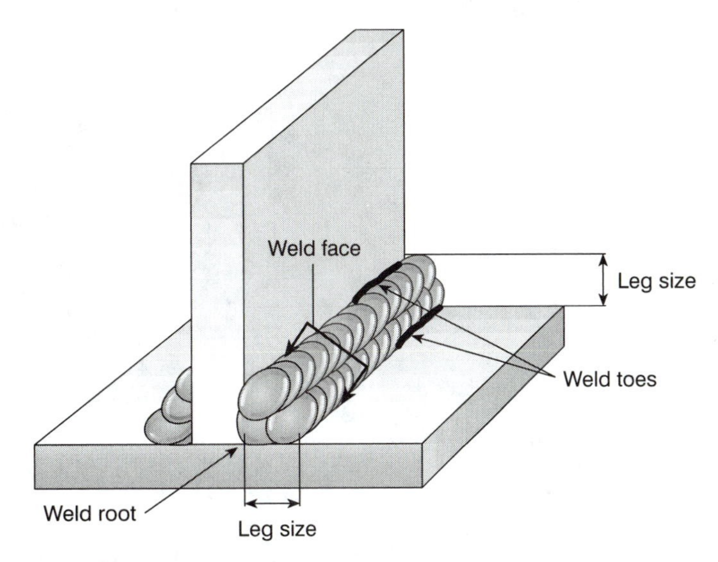

FIGURE 7–3 The parts of the fillet weld.

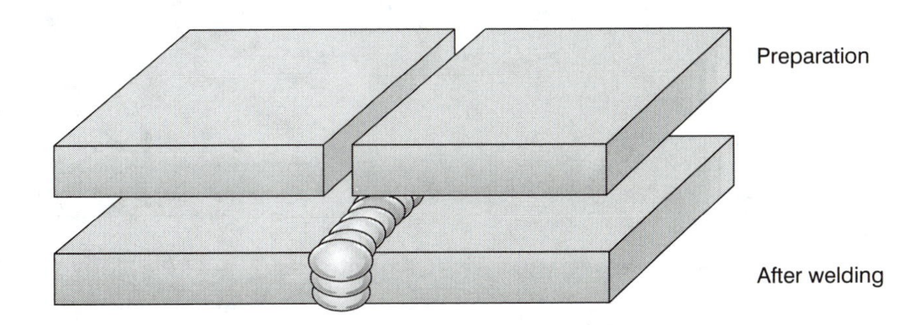

FIGURE 7–4 Square-groove weld.

to the back side of the joint. A groove weld can take the form of one of the eight designs for the butt joint. This book will examine three of these designs, the square-groove weld, the bevel-groove weld, and the V-groove weld, with specifications from the AWS *Structural Welding Code—Steel D1.1.*

The **square-groove weld** is for use in the butt joint without edge preparation. The square-groove weld is effective for joining sheet metal and plate with limited edge preparation. The corner joint is one application of the square-groove weld. The butt joint is another. For shielded metal arc welding, the maximum recommended base metal thickness is ¼ inch for complete root penetration when making a square-groove weld from both sides of a butt joint, as shown in Figure 7–4. Other factors that have to be considered in welding plate with ¼ inch thickness are the choice of electrode, the welding position, amperage setting, and the **root opening.**

The butt joint has to be prepared for welding a V-groove weld, or any of the other groove welds that require a **bevel,** or cutting away of some of the edge. The electric grinder, oxyacetylene torch, and mechanical beveler are three tools that can be used to

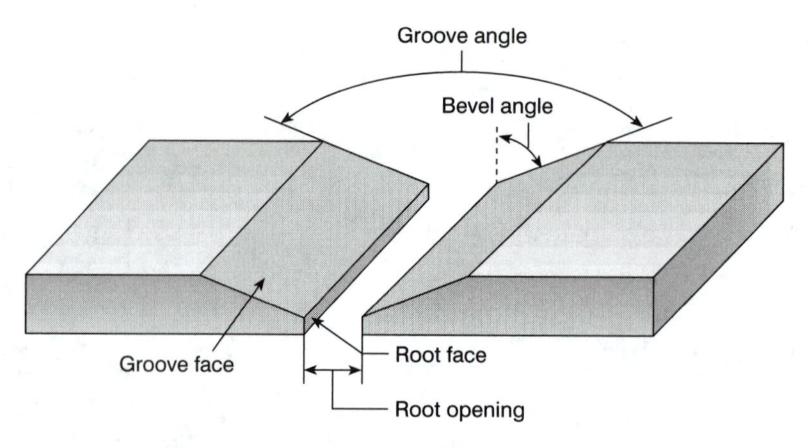

FIGURE 7–5 The parts of the V-groove butt joint.

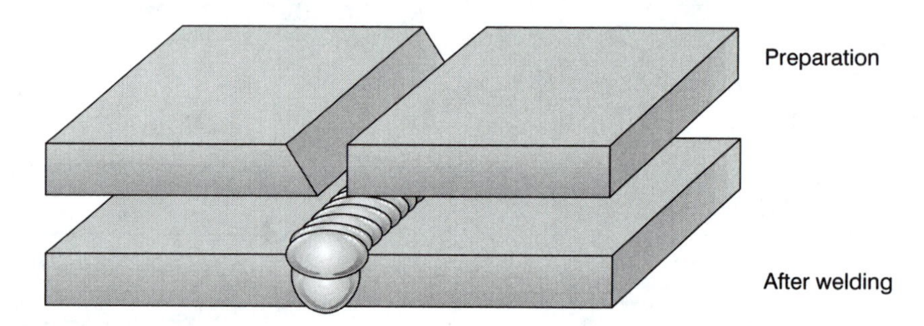

FIGURE 7–6 Bevel-groove weld.

bevel plate. The purpose of beveling plate is to aid root penetration into the joint. After preparation, the parts of one butt joint look like Figure 7–5. The **groove angle** is formed when two bevels form a butt joint. The **groove face** is the surface where the pieces of the joint meet. The root opening is the space between the pieces that form the joint. The **root face** is the edge that is not removed by beveling. Sometimes a root face is used to prevent melt-thru on the first pass, which would cause too much of the base metal to melt away.

The **bevel-groove** weld is a weld in which the edge of one piece forming the joint is beveled. The bevel-groove weld is effective when preparing only one piece of the joint, as in Figure 7–6. For shielded metal arc welding, when welding from one side with **backing material,** a 45° bevel angle with a root opening of ⅜ inch is recommended.

The **V-groove weld** is a weld in which the edges of both pieces forming the joint are beveled, as shown in Figure 7–7. The V-groove weld is an effective weld when both pieces of the joint must be prepared. For shielded metal arc welding, the maximum recommended base metal thickness for complete joint penetration is unlimited. When welding from one side with backing material, the degree of the groove (two bevels) decides the size of the

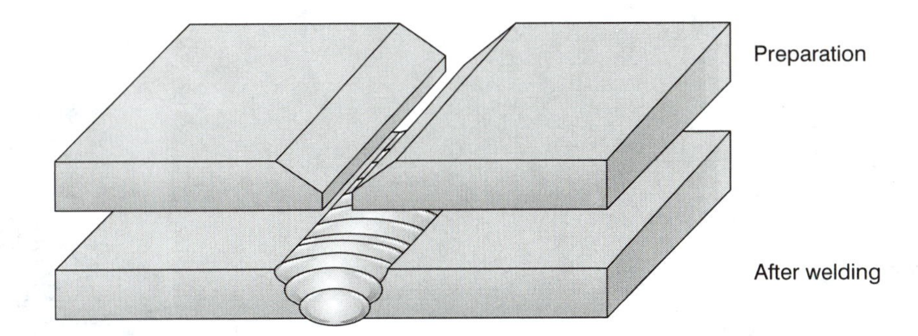

FIGURE 7–7 V-groove weld.

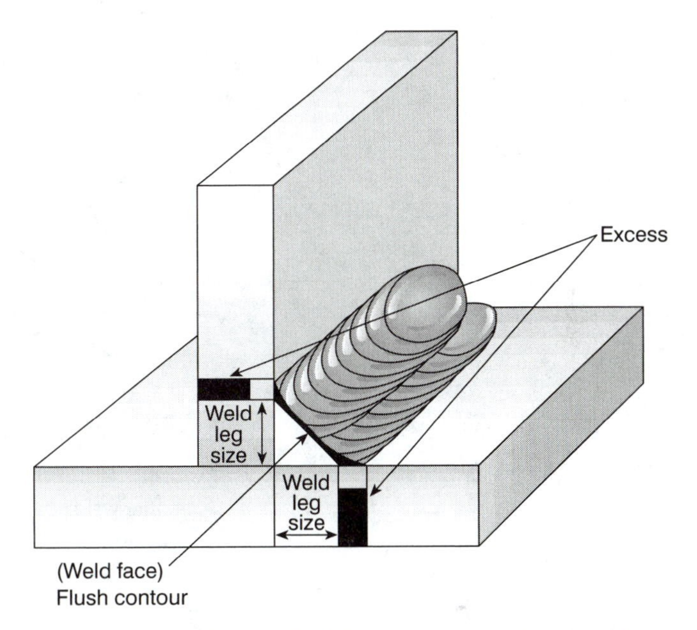

FIGURE 7–8 Measured size of a fillet weld is the leg size from the point of the flush contour.

root opening: a 45° groove angle takes a root opening of ¼ inch, a 30° groove angle requires a root opening of ⅜ inch, and a 20° groove angle needs a root opening of ½ inch.

Many weldments fail because only one side of the joint was welded. Observe how easy it is to break a weld by applying the force of a hammer on the side opposite the weld. One way of avoiding this potential problem is to weld both sides of the joint or use groove welds with complete joint root penetration.

Weld Size

The **fillet weld legs** help to establish the size of the fillet weld. Each fillet weld has two legs and two **weld toes.** The size of the fillet weld is measured off a straight line connecting each leg at the point of a **flush contour,** extending to the **joint root,** as pictured in Figure 7–8. The joint root is the point where the surfaces of the base metal are closest. The contour of the weld face, whether it is

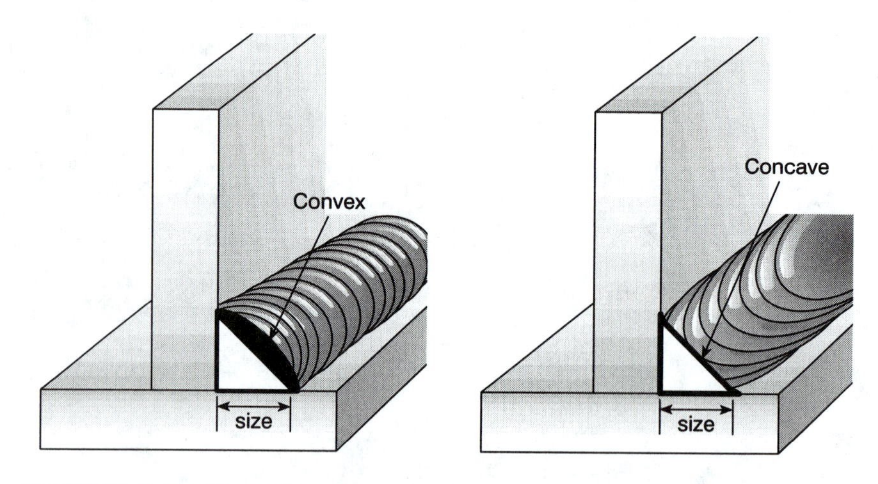

FIGURE 7–9 Profiles of convex weld face and concave weld face.

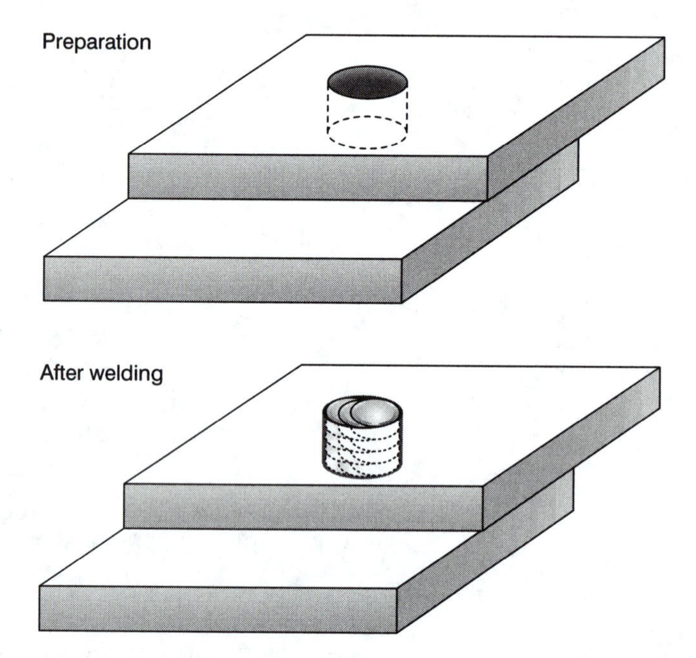

FIGURE 7–10 Plug weld in circular hole.

convex or **concave,** is not part of figuring the size of the weld (see Figure 7–9).

Other Welds

The American Welding Society lists 19 different welds that are used for joining metal together. While the fillet weld and the groove welds receive most of the attention, there are some other welds to become familiar with. The **plug weld** is considered a remnant of the riveting age. The plug weld is a weld in a circular hole of one piece joined with a second piece in a lap joint (Figure 7–10). The size of the plug weld is given by the diameter of the hole. The hole can be beveled for accessibility or increased hold-

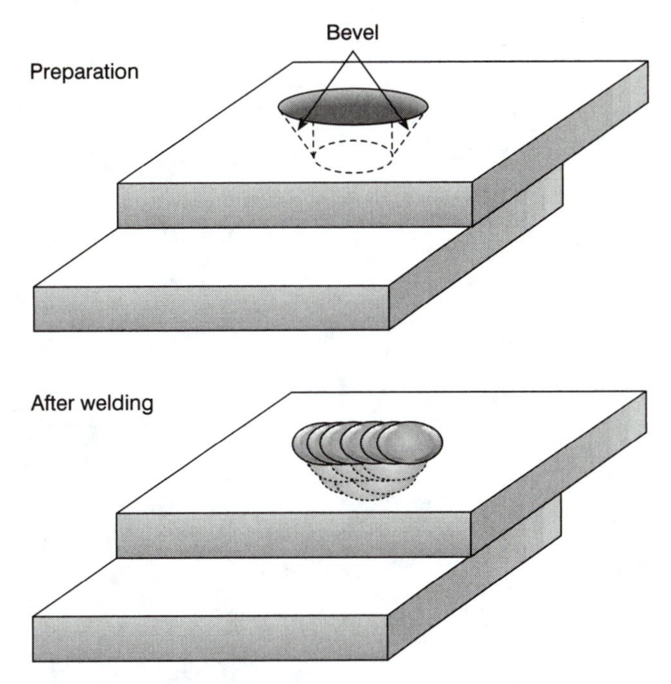

FIGURE 7–11 Plug weld in beveled hole.

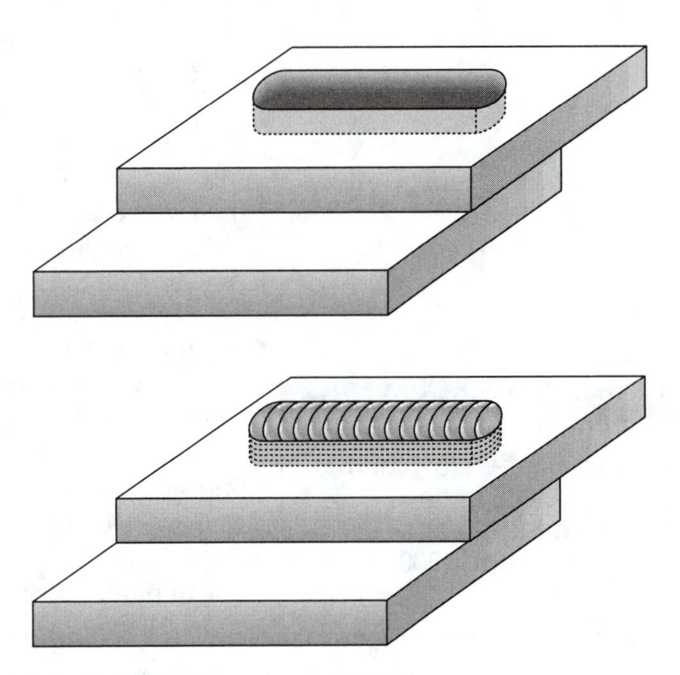

FIGURE 7–12 Slot weld in a elongated hole.

ing strength, when weldment is subjected to wear, as in Figure 7–11, and sometimes it is only partially filled with weld. The plug weld is used in applications in which welding along the edge of a joint is not enough or is not practical.

The **slot weld** is another remnant of the riveting age. The slot weld is an elongated plug weld, (Figure 7–12). The slot weld can also be beveled for accessibility and increased holding strength.

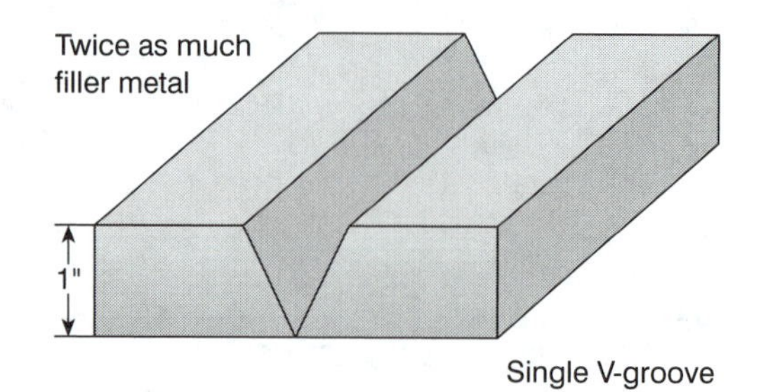

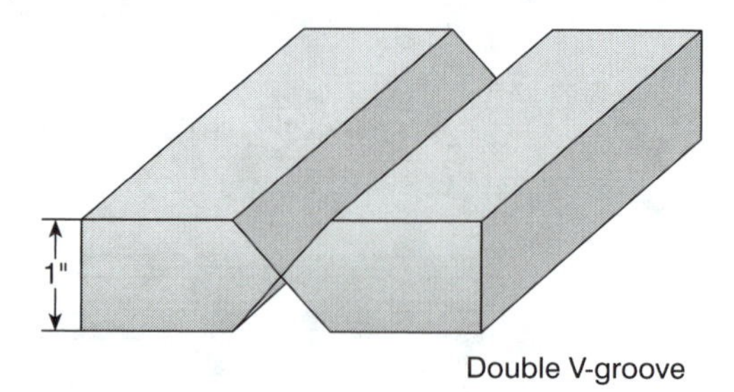

FIGURE 7–13 Double-groove weld versus single-groove weld; note that the single-groove weld will require more welding time.

Selecting a Joint Design

Selecting a joint design for welding depends on several factors. One objective is to determine the use of and the forces acting upon the joint. A second objective is to choose joints that involve the least amount of welding and the least amount of preparation before welding. A third objective is that thinner metals, the use of smaller root openings, and smaller bevel angles all require less filler metal. A fourth objective considers that, for thick plate, double-groove welds are preferred over single-groove welds (Figure 7–13). Double groove welds require less filler metal, which saves time and money.

Sizing welds is very important. Remember that the more welding done and the more heat and the more force pulling at the joint, the more **distortion.** Other ways of eliminating unnecessary welding can cut expenses and make a company more competitive.

Since overwelding causes undo stress on the joint and adds time and money to the job, every effort should be made to size each weld to meet the requirements of the weldment. The maximum size weld should equal the thickness of the thinner member in the joint. The minimum size weld depends on the application, taking into account those physical forces that will be acting on the weldment.

ARC FLASH

An 1/8 inch fillet weld 1 inch long using 60,000 series electrodes has a shear strength of around 1500 pounds.

SAFETY REMINDER

Weld failure can have devastating consequences. Be sure your welding is never in question.

At this stage in your training, limit yourself to welding projects where the engineering problems have been solved by others. Taking an advanced welding course can then be a rewarding experience that will benefit you the rest of your life.

However, always keep in mind that rules and common practices do have their limitations. Physical forces can have extraordinary effects on a weldment. Consequently, limit welding to class projects in which weld failure will not cause injury to yourself or others.

REVIEW

1. Why is the fillet weld more popular than the groove weld?
2. What is the purpose for beveling a joint before welding?
3. Which joint, the bevel-groove weld or the V-groove weld, requires more preparation?
4. What are the results of overwelding?
5. How is the size of the fillet weld measured?

Lab Exercises

1. How many of the five joint designs can you locate in the shop?
2. How many of the following welds can you find in the shop: (a) fillet weld, (b) square-groove weld, (c) bevel-groove weld, (d) V-groove weld.
3. Locate fillet welds on structural beams around the shop. Then try to determine their sizes.

REVIEW QUESTIONS FOR UNIT 7

Multiple Choice

Choose the best answer.

1. Which is one of the five basic joints?
   a. Square-groove
   b. Edge
   c. Lap
   d. Butt
   e. a, b, and c
   f. b, c, and d
2. An important consideration in joint design is:
   a. Maximum strength with efficiency.
   b. As many different types of joints as possible.
   c. Maximum number of welds possible.
   d. Least amount of welding.
   e. a and c.
   f. a and d.
3. Parts of a fillet weld include:
   a. Weld toes and weld root.
   b. Weld legs and size.
   c. Weld face and weld length.
   d. a and b.
   e. a and c.

4. Parts of butt joint include
   a. Groove face and bevel angle.
   b. Root face and root opening.
   c. Groove angle and groove face.
   d. All of the above.
   e. a and b.
5. Methods for beveling plate include:
   a. Electric wrench.
   b. Oxyacetylene torch.
   c. Mechanical beveler.
   d. Heavy duty sander.
   e. a and b.
   f. b and c.
6. Which requires the least amount of weld?
   a. Square-groove weld.
   b. Bevel-groove weld.
   c. V-groove weld.
   d. a and b.
   e. None of the above.
7. The term *concave* is associated with the:
   a. Weld root.
   b. Weld face.
   c. Weld toes.
   d. Joint root.
   e. None of the above.
8. Remnant(s) of the riveting age include the:
   a. Groove weld.
   b. Plug weld.
   c. Fillet weld.
   d. a and b.
   e. b and c.
9. A double-groove weld:
   a. Is used on sheet steel.
   b. Uses more filler metal than single-groove welds.
   c. Is used on thick plate.
   d. Takes more time and costs more money.
   e. None of the above.
10. Overwelding:
    a. Causes under stress on the joint.
    b. Costs more money.
    c. Makes for more force pulling on the joint.
    d. a and c.
    e. a, b, and c.

Short Answer

1. What are the five basic joints?
2. Give the parts of a fillet weld.
3. When is complete root penetration achieved?
4. How is the root face different from the root opening?
5. What do fillet weld legs tell you?
6. If the fillet weld legs are the same, which requires more filler metal, a convex or concave weld face?

7. What are three results of overwelding?
8. Which of the following welds requires more preparation, and why: square-groove weld, bevel-groove weld, or V-groove weld?
9. Why would a corner joint design work better than a tee joint in fabricating a container to hold a liquid?
10. Which joint is most often used with a V-groove weld?

ADDITIONAL ACTIVITIES

Matching

Match each of the five basic joints with its identifying characteristics.

1. Tee joint ____________
2. Lap joint ____________
3. Edge joint ____________
4. Butt joint ____________
5. Corner joint ____________

a. backing material
b. fillet weld
c. plug weld
d. thin metal to limit distortion
e. preparation over ¼ inch for complete root penetration
f. requires the most skill and preparation
g. joining plates at 90° to each other
h. horizontal position most common for welding
i. two pipes joined together, end to end
j. two pipes joined, telescoping out of the each other
k. groove weld used most often
l. spot weld

Writing Skills

1. Write a paragraph to explain what should be considered in joint design.
2. Write a paragraph using the following terms in sentences: *weld-ment, concave, lap joint, bevel, fillet weld, fillet weld legs, ¼-inch thick, weld face.*

Math

The job calls for fabricating a tank out of ⅛-inch sheet steel. Remember that the volume of a rectangular container = width × length × height.

1. How many full sheets of 4' x 8' are needed to construct the tank shown in Figure 7–14, including one 3' × 2 ½' baffle?
2. If a press brake is available to bend up the sides to reduce welding and limit distortion, the total length of welding will be reduced to ____________.
3. What is the volume of the tank?
4. Convert your answer from Question 3 to gallons. Remember: one gallon = 231 cubic inches, and 12" × 12" × 12" = 1728 cubic inches, or 1 cubic foot.

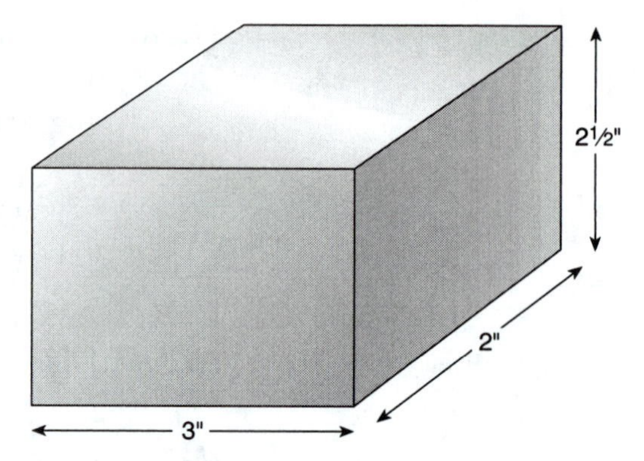

FIGURE 7–14 Tank to be constructed.

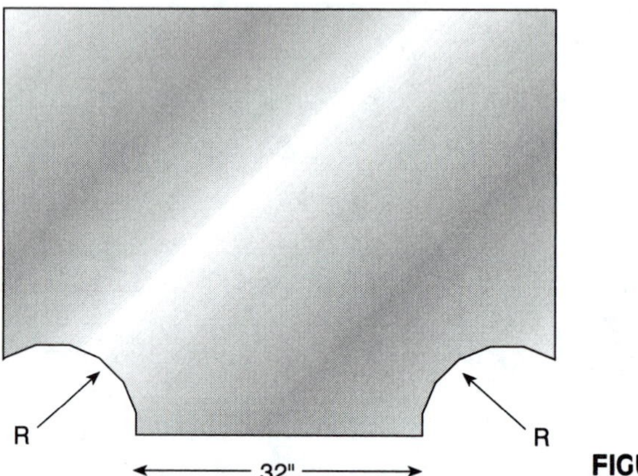

FIGURE 7–15 Baffle.

5. What is the radius for the cutouts in the baffle for the tank, shown in Figure 7–15?

Puzzlers

1. What determines a decision to change from a square-groove weld to a bevel-groove weld?
2. What is to be considered in deciding on a joint design?

UNIT 8

Welding Electrodes

1. APPLICATION OF ELECTRODES

Purpose of Electrodes

The electrode has a purpose of conducting electricity to the point of welding. At this point a short circuit is created, and intense heat is generated (Figure 8–1). The force of attraction carries molten metal across the arc to the weld pool, where filler metal mixes with the base metal.

The weld, functioning like a screw or bolt, becoming a fastener for holding metal together. Welds have become the replacements for one kind of fastener, the rivet. Rivets were readily used at one time in massive metal structures like bridges. Welded designs have lightened the load requirements over riveted construction. For bridges, this means more strength in the weight of the steel can be applied to supporting more cars, trucks, buses, and trains that travel overhead.

FIGURE 8–1 Intense heat and light are produced by welding.

Types of Electrodes

As mentioned in Chapter 3, the first two digits (E**70**18)—or the first three digits if a five digit number (E**100**18)—refer to the minimum **tensile strength** of the weld in pounds per square inch (**70** × 1000 = 70,000 or **100** × 1000 = 100,000). The third digit (E70**1**8) provides information on welding position: a number 1 meaning the electrode can be used in all four welding positions, a number 2 (as in E70**2**4) meaning the electrode is recommended for use in the flat and horizontal positions only.

The fourth digit (E701**8**) gives the choice of welding current, information on the flux covering, and characteristics of the weld bead and slag formation (see Table 8–1). The table shows that while some electrodes are designed for both AC and DC welding, many electrodes are designed for use only with DC welding.

TABLE 8–1 Characteristics of common electrodes.

E. Number	*Type of covering*	*Position for Welding*	*Current Available*	*Penetration*	*Shape of Weld Bead*	*Slag*	*Slag Removal*	*Minimum tensile strength*
E6010	Cellulose sodium	All positions	DCEP	Deep	Concave	Thin	Somewhat easy	62,000 psi
E6011	Cellulose potassium	All positions	AC DCEP DCEN	Deep	Concave	Thin	Somewhat easy	62,000 psi
E6012*	Titania sodium	All positions	AC DCEN	Medium	Convex	Moderate	Easy	67,000 psi
E6013*	Titania potassium	All positions	AC DCEP DDEN	Shallow	Convex	Moderate	Easy	67,000 psi
E7014*	Iron powder titania	All positions	AC DCEP DCEN	Medium	Flat to convex	Heavy	Easy	70,000 psi
E7015**	Low hydrogen sodium	All positions	DCEP	Medium	Convex	Moderate	Somewhat easy	70,000 psi
E7016**	Low hydrogen potassium	All positions	AC DCEP	Medium	Convex	Moderate	Very easy	70,000 psi
E7018**	Iron powder low hydrogen	All positions	AC DCEP	Medium	Convex	Heavy	Very easy	70,000 psi
E7024	Iron powder titania	Flat and horizontal	AC DCEP DCEN	Medium	Flat	Heavy	Easy	70,000 psi

*Follow the manufacturer's recommendations for oven storage. Suggested oven temperature is from 20° to 40°F; ANSI/AWS A5.1.
**Follow the manufacturer's recommendations for oven storage and drying. Suggested oven holding temperature is from 50° to 250°F; ANSI/AWS A5.1.

Low-hydrogen electrodes are used in place of a E6010, E6011, or E6012 in many applications. These applications include earth-moving equipment such as graders or bulldozers, the boat-building industry, and building construction. Low-hydrogen electrodes are designed to resist moisture absorption for welding carbon steels and low-alloy steels. They prevent **underbead cracking** in some types of steels that can result when hydrogen is absorbed into the weld.

A popular low-hydrogen electrode is E7018. This electrode has application in welding some types of steel where there is a danger of underbead cracking. The slag created by E7018 is easily removed. Generally, E7018 should not be used on steel that has not been cleaned nor in open root joint designs.

Although mild steel and low-alloy electrodes are very popular, other varieties are used, too. Stainless steel electrodes are used to arc weld stainless steel and for special applications. Nickel alloy electrodes are used to arc weld nickel alloys and cast iron. Aluminum electrodes are used to arc weld aluminum, and copper alloy electrodes are used to arc weld copper alloys. **Hard-surfacing** electrodes are used to build up and overlay metal that is subject to wearing away by abrasion.

The numbers put on stainless steel electrodes have meaning. Consider the following:

E308-15
E308-16
E308L-15 & 16

The letter **E** for electrode is followed by three digits, in this case **308,** which stands for the type of stainless steel in the filler metal. The two digit number **15** indicates the choice of welding current should be DCEP. The two digit number **16** indicates the choice of welding current should be DCEP or AC. Extra low-carbon stainless steels usually require electrodes with the letter **L** to prevent a reduction in corrosion resistance. With 308H electrodes the letter **H** means high-carbon, and with 308-HC the letters **HC** means high carbon following under military specifications. Other letters such as **Mo** (molybdenum), **Cb** (columbium), and **Ni** (nickel) are also used as indications of other elements added to the electrode for prevention of corrosion and cracking, as well as added ductility and strength.

ARC FLASH

Care must be taken in the storage of electrodes. Always follow the manufacturer's recommendations.

Other welding processes do a better job of welding aluminum than shielded metal arc welding. However, sometimes small projects can be completed using this process. There are two types of aluminum covered electrodes. These electrodes are designed for welding using DCEP.

Al-2 & Al-43

While some aluminum can be welded with covered electrodes by shielded metal arc welding, always run a test procedure to find out for sure. The key to welding aluminum is to clean and preheat the base metal before striking an arc. Preheating may be required to help assure that the electrode readily melts together with the base metal. The electrodes themselves should be moisture free, coming from sealed packaging or heated for one hour at 400° F. Moisture will cause the flux covering to flake off quite easily.

There are always going to be problems that have to be solved before the welding can be completed. Be sure the electrode filler metal selected for the project is not one of the problems. Purchase electrodes from a reputable supplier.

Choice of Electrodes

The choice of an electrode for a given application can depend on several things. The size of the electrode, the amperage setting, the welding position, the composition of the base metal, the thickness, shape, and fit-up of the joint, are some of the factors to consider in choice of an electrode. The choice of an electrode answers some important questions. If welding problems occur after beginning a project, refer to the troubleshooting guide in Chapter 5.

The size of the electrode is measured by the diameter of the bare wire. The flux covering is not used in measuring size because the thickness of the flux varies among the different electrodes (see Figure 8–2). Size is important because a larger diameter electrode requires a higher amperage setting on the power source (Figure 8–3). A higher amperage setting results in more filler metal being deposited into the joint by a single weld bead. But higher amperage can also limit the welding positions because gravity affects the ability to control the weld pool. A smaller weld pool of lower amperage is affected to a lesser degree by gravity. A smaller weld pool is less likely to flow out of the joint in the vertical position and the overhead position.

With experience, setting the amperage for a given electrode will not be a mystery. If the welder matches the amperage on the power source (within the approximate range provided in Table 8–2), the possibility of melt-thru can be minimized. Melt-thru is a

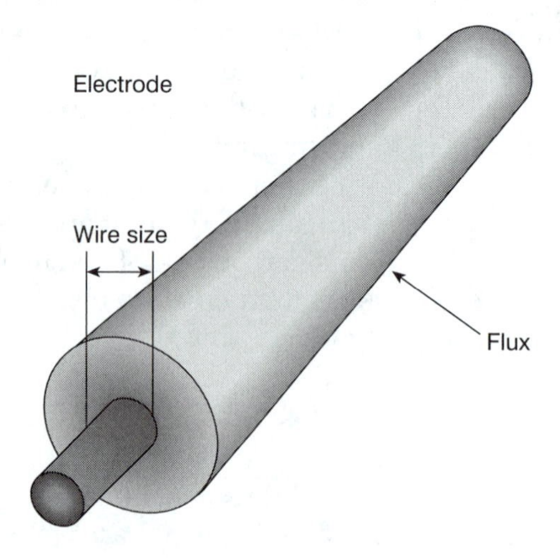

FIGURE 8–2 The size of electrodes is taken from the wire size.

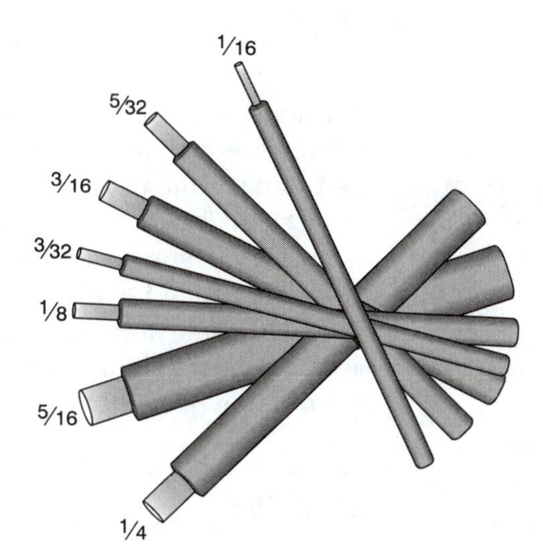

FIGURE 8–3 Electrodes of various sizes from 1/16-inch to 5/16-inch.

TABLE 8–2 Range of settings for different size electrodes.

Diameter of Electrode (inches)	*Amperage*	*Voltage*
1/16	20–40	17–20
5/64	25–60	17–21
3/32	40–120	17–21
1/8	55–170	18–22
5/32	80–270	18–22
3/16	140–330	20–24
7/32	200–400	20–25
1/4	250–500	20–26
5/16	275–525	22–27

problem for the inexperienced welder working with thin steel. And thin rusty steel that has been cleaned by grinding becomes even thinner. The lower amperage of smaller diameter electrodes is one solution for welding thin steel. The movement of the electrode, starting and stopping the arc at short intervals, may also help; this technique can be used to build a foundation of filler metal around holes before filling them in. Shorter length weld beads do not create the excessive heat of longer weld beads. So careful observation coupled with experience should solve most welding problems for setting the amperage and eliminating the problem of melt-thru.

Successfully completing the exercises provided in this book for the flat position and the horizontal position is preparation for welding many projects. To perform out-of-position welding in the vertical position and overhead position, lower the amperage and try using a smaller diameter electrode. A smaller size weld pool gives better control to help counteract the effects of gravity.

While most steels can be welded successfully, some steels cannot be welded without difficulty or without running up a prohibitive cost. Many problems can be avoided by limiting the welding to base metals with low-carbon steel composition. These steels are readily welded by shielded metal arc welding. Unless fabricating a structure which will be subjected to undo stress that requires a specialty steel, the welding electrodes covered by Table 8–1 will handle most welding applications.

Do not let thickness and fit-up of the joint affect the quality of the work. Try to stay away from projects involving material over ¾ inch thick that may involve preheating procedures. Never be afraid to spend the time necessary to make close fitting joints that will require less welding. Taking the time to prepare for welding will aid in the production of quality work that any welder will be proud of.

ARC FLASH

Nickel, copper nickel, cast iron, aluminum, and stainless steel are other electrodes available from dealers in welding supplies.

REVIEW

1. How is a weld like a screw or bolt?
2. What are a few techniques for welding on thin metal?
3. What does the 1 in E7018 mean?

Lab Exercises

1. a. Make a list of the types of electrodes used in the shop.
   b. Now gather as much information as you can about each type.
2. Write a description of the differences that you can observe between them.
3. a. Identify any cast iron, aluminum, or stainless steel electrodes in the shop.
   b. Write a description of the particular electrode.
4. Does the manufacturer of these cast iron, aluminum, or stainless steel electrodes require anything special in the way of storage?

2. THE LIST OF A FEW MANUFACTURERS

Brands of Common Electrodes

Table 8–1 shows several types of electrodes on the market. A few electrode manufacturers are given in Table 8–3. Do not be taken in by advertisements about high priced specialty electrodes with catchy names that are supposedly made for unique welding situations. While a few types of electrodes within the American Welding Society's number classification system are designed for particular applications, many mild steel fabrication projects can be completed with a box of ⅛ inch E6011s. E6011 even works well on carbon steel that cannot be cleaned sufficiently.

If a power source has both AC and DC capabilities, be sure to buy electrodes that can be used for welding both AC and DCEP. Whatever brand is used, be sure to read the manufacturer's information on each of the electrodes. Whenever trying a new brand, even from the same manufacturer, ask to try out a handful before buying in quantity. Electrodes are not all the same.

TABLE 8–3 Manufacturers' brand names for electrodes.

	AWS Number (Mild Steel)		
Manufacturer	**E6010**	**E6011**	**E6012**
McKay	6010	6011	6012
Weldrite	E6010	E6011	E6012
Hobart	Pipemaster 60 Hobart 60AP	Hobart 335A Hobart 335C	Hobart 12 Hobart 12A
Alloy Rods	AP100 SW 610	SW-14	SW 612
Lincoln	Fleetweld 5 Fleetweld 5P	Fleetweld 35 Fleetweld 35LS Fleetweld 180	Fleetweld 7
	Mild Steel		
	E6013	**E7014**	**E7024**
McKay	6013	7014	7024
Weldrite	E6013	E7014	E7024
Hobart	Hobart 447A Hobart 447C	Hobart 14A	Hobart 24 Hobart 24-1
Alloy Rods	SW-15 6013LV	SW-15IP	7024 SW-44
Lincoln	Fleetweld 37 Fleetweld 57	Fleetweld 47	Jetweld 1 Jetweld 3
	Mild Steel Low-Hydrogen		
	E7016	**E7018**	**E7028**
McKay	7016	7018 7018 XLM 7018 M 7018-1*	
Weldrite		E7018	
Hobart	Hobart 716	Hobart 418 Hobart 718 Hobart 718C*	Hobart 728
Alloy Rods	70LA-2	7018 7018AC 7018-1*	
Lincoln		Jetweld LH-70*	Jetweld LH-3800

SAFETY REMINDER

Never leave an electrode in its holder unattended. Avoid the possibility of creating a potentially dangerous situation should the electrode arc.

Care of Electrodes

Keep electrodes in a dry place to prevent the formation of rust on the wire core and moisture in the flux. It would be frustrating to begin welding and watch the flux covering drop off the electrode. Imagine how the quality of the weld would be lowered. Always give low-hydrogen electrodes the added treatment of storage in an oven. While electrodes that become wet can be reconditioned following the manufacturer's recommendations, the welding procedure may not allow the exercise of this practice. Purchase brand name electrodes and store them properly to maintain quality standards in your work.

LAB EXERCISES

1. a. Who manufactures the electrodes used in your shop?
   b. What are their brand names?
2. Make a list of the brand names from other manufacturers that may be equivalent.
3. What are the brand names of any electrodes in the shop that can be used for:
   a. Only DCEP?
   b. Both DCEP and AC?
4. Sketch out a project, describing the joint designs you will be using.

REVIEW QUESTION FOR UNIT 8

Multiple Choice

Choose the best answer.

1. Welds:
   a. Are replacements for rivets.
   b. Join material together.
   c. Lighten load requirements in design.
   d. All of the above.
   e. a and b.
2. E7018 refers to:
   a. Low-hydrogen.
   b. DCEN.
   c. 60,000 psi.
   d. Flat, horizontal, vertical, and overhead positions.
   e. a and d.
3. E6011 means:
   a. Only AC.
   b. Cellulose sodium covering.
   c. Shallow penetration.
   d. Heavy slag.
   e. None of the above.
4. E7024 tells you:
   a. Low-hydrogen.
   b. Flat position only.

c. Heavy slag.
d. DCEP only.
e. None of the above.

5. Low-hydrogen electrodes:
   a. Are designed to resist moisture absorption.
   b. Prevent underbead cracking.
   c. Are recommended for oven storage.
   d. Are for welding low-alloy steels.
   e. All of the above.
6. Hard-surfacing electrodes:
   a. Are used for buildup.
   b. Are designed to slow wearing away.
   c. Are used with soft metal.
   d. a and c.
   e. a and b.
7. The following factors influence your choice of electrode:
   a. Welding position.
   b. Composition of the base metal.
   c. Fit-up of the joint.
   d. All of the above.
   e. None of the above.
8. A higher amperage:
   a. Forms a smaller weld pool.
   b. Can limit the welding position.
   c. Deposits more material.
   d. a and b.
   e. b and c.
9. A lower amperage:
   a. Requires a smaller electrode.
   b. Makes melt-thru more likely.
   c. Requires a larger electrode.
   d. All of the above.
   e. None of the above.
10. A close fitting joint:
    a. Requires more welding.
    b. Requires less welding.
    c. Requires more time in preparation.
    d. a and b.
    e. b and c.

Short Answer

1. What are welds?
2. What does the 70 in 7018 stand for?
3. What does the 1 in 7018 stand for?
4. What does the fourth digit of a four-digit electrode, or the fifth digit of a five-digit electrode, stand for?
5. Define *slag?*
6. Give the welding positions for E7024?
7. How does the diameter or size of the electrode affect the welding time?
8. How is the size of the electrode measured?

9. Why use a smaller electrode for vertical position or overhead position?

10. What is the smallest electrode available?

ADDITIONAL ACTIVITIES

Fill in the Blanks

1. ______________ electrodes work well on steel that cannot be cleaned properly.

2. ______________ electrodes are used in situations where underbead cracking would be a problem.

3. ______________ electrodes are the choice for building up metal subject to abrasion.

4. E7018 electrodes require ______________ or ______________ current.

5. The number ______________ means the current for the______________ steel electrode should be DCEP.

6. The number ______________ for the same type of electrode in question 5 means the choice of current can be DCEP or______________ .

7. Setting the ______________ should not be a mystery.

8. A ______________ amperage means ______________ weld pool.

9. On the other hand, a ______________ amperage means a ______________ weld pool.

Writing Skills

1. Write a paragraph about the different factors that go into choosing an electrode.

2. Give an explanation of when to use an E6011 electrode and when to use an E7018.

3. Write sentences that compare the different characteristics of each of three electrodes used in the shop.

Math

About *circles and cylinders:*

d = diameter r = radius

c = circumference h = height

π = Greek letter pi

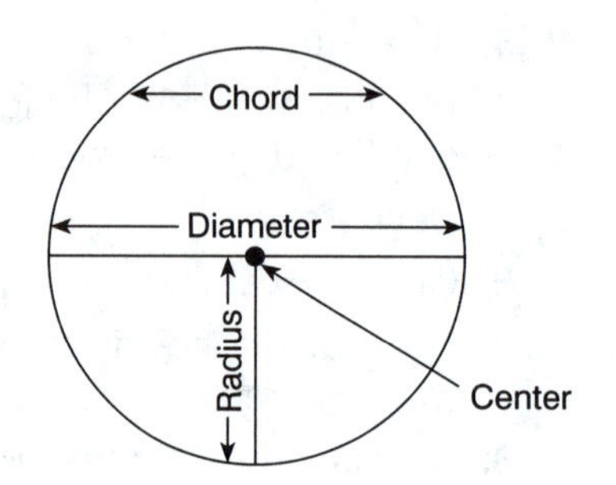

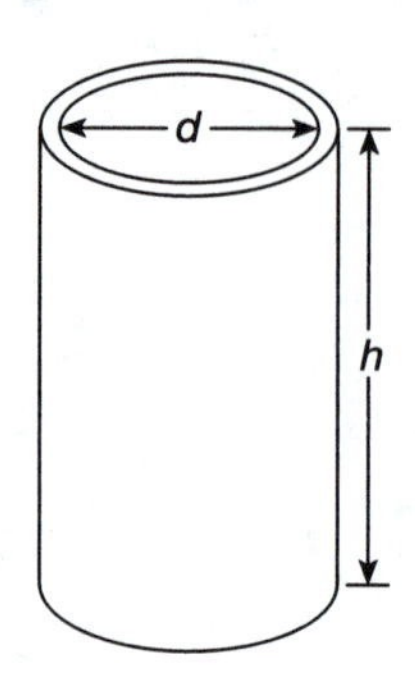

Formulas:

circumference = $\pi \cdot d$

r^2 (radius multiplied by itself) = $r \cdot r$

area of circle = $\pi \cdot r^2$

volume of cylinder = $(\pi \cdot r^2) \cdot h$

diameter = 2r $\pi = 3.14$

Find the volume of the following cylinder. Answer in cubic feet and inches.

	cylinder			
	1	*2*	*3*	*4*
diameter	2'	30"	2'7"	3½'
height	3'	25"	4'2"	5½'

1. ____________ **2.** ____________

3. ____________ **4.** ____________

Solve using the same numbers as for problems 1–4, only think that for the following problems the diameters are outside diameters for steel that is ⅛-inch thick.

5. ____________ **6.** ____________

7. ____________ **8.** ____________

A chord is a straight line that intersects a circle at two points not going through the center. To locate the center of a circle, divide two chords in half (bisect them) and construct a right angle at the point of bisection of each.

9.

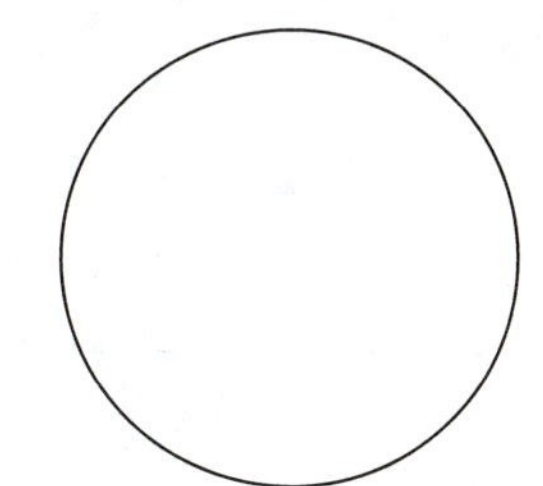

10.

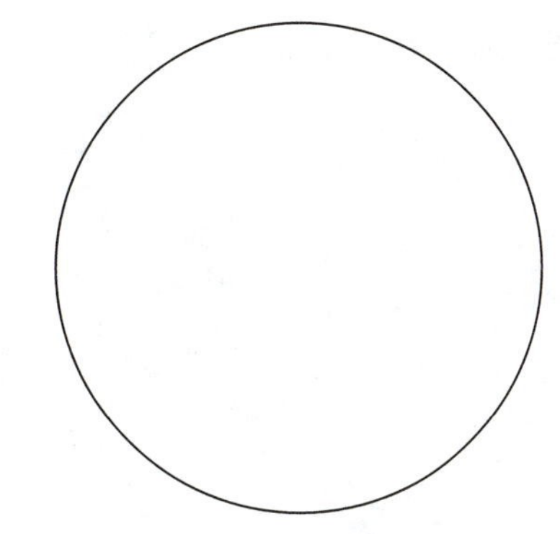

Puzzlers

1. There is no choice, the job will have to be welded using E7018. Unfortunately, an ⅛-inch gap will have to be filled. The good news is that the welding can be completed from either the front or backside of the joint. What can be done to assure that the first pass turns out to be a quality weld?
2. You begin welding only to discover that the flux covering is breaking off the electrode as it heats up. What might be the problem, and what can be done as far as a possible solution might go?

UNIT 9

Heat and Distortion

1. EFFECTS OF WELDING

As stated in an earlier unit, steels are selected for their properties. Medium-carbon steels are used in applications requiring moderate hardness and wear resistance. Low-alloy steels are used in applications subject to a higher degree of toughness and hardness with greater tensile strength than low-carbon steels. Medium-carbon and low-alloy steels are used in industries such as mining, agriculture, and shipbuilding.

Both medium-carbon and low-alloy steels require precautions in welding. The concern is that **martensite** might form in the heat-affected-zone (HAZ), which can lead to cracking, and can make preheating and/or postheating necessary. Martensite is a hard pin-like **grain structure** that forms when some steels cool too rapidly. Martensite forms in steel when the cooling rate of a welded joint is so fast that a more **ductile** grain structure is unable to form.

When preheating is necessary, the preheat temperature of the base metal for welding is determined by the type of steel it is, among other factors. This temperature can range from 150° to 550° F; such a temperature is practical without requiring specialized equipment. Preheating helps to slow the rate of cooling to prevent the formation of martensite.

Expansion and Contraction

All metals are affected by heat differently. One difference is in the way metals expand when heated. This expansion, which varies with the type of metal (aluminum versus steel for example), can be measured and is called the **coefficient of linear expansion** (see Table 9–1). Note in that table that steel will expand at only about one-half the rate of aluminum. This means that aluminum is

Table 9–1 Expansion coefficient for common metals per rise in one degree of temperature.

Metal	*Expansion Coefficient per Inch*
Aluminum	0.0000123
Copper	0.0000088
Steel	0.0000063
Cast Iron	0.0000056
Stainless Steel	0.000009

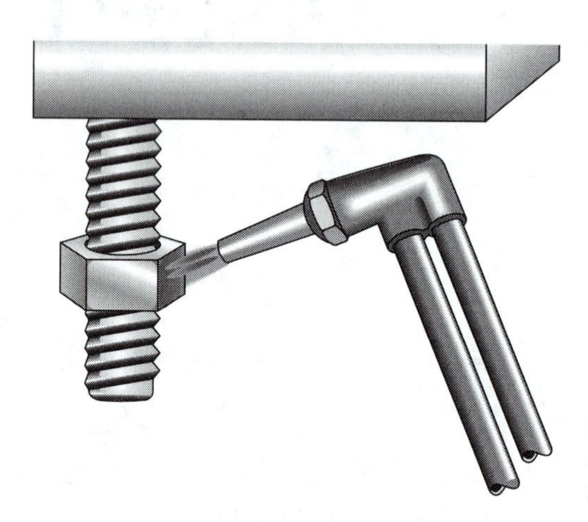

FIGURE 9–1 Heat expands metal, making removal of a nut easier.

nearly twice as sensitive to heat and more readily subject to distortion than steel. A person can observe the effects of this expansion when hearing a metal barrel pop on a sunny day or your oven make a noise as it cools down. Knowledge of how heat causes expansion can be used to the welder's advantage. Applying the flame of a torch to expand the nut on a rusted bolt so it can be twisted off without difficulty (Figure 9–1) is one application of controlled heating, making it easy on both the tools and a person's back. Obviously, on large welding projects the measurements will change depending on the temperature at a given time of day.

While metal expands when heated, metal also contracts on cooling. If metal is heated evenly and allowed to cool evenly, it will return to its original shape if not restricted. As an example of restriction, put a bar 1 inch diameter by 6 inches long into a vice and heat evenly to orange red, then let the bar cool. The bar having been restricted should fall from the vice upon cooling, thicker but shorter than before, as in Figure 9–2. This bar becomes thicker due to **distortion.** In welding, metal never returns to its original shape because heating is always uneven as shown by the example of the bar in a vice. A welder quickly learns how heat can cause the distortion of metal by attempting to weld a frame together without tack welds in place, as shown in Figure 9–3. The frame may move out of position, so distorted that it becomes impossible to bring it back together without cutting out the welds.

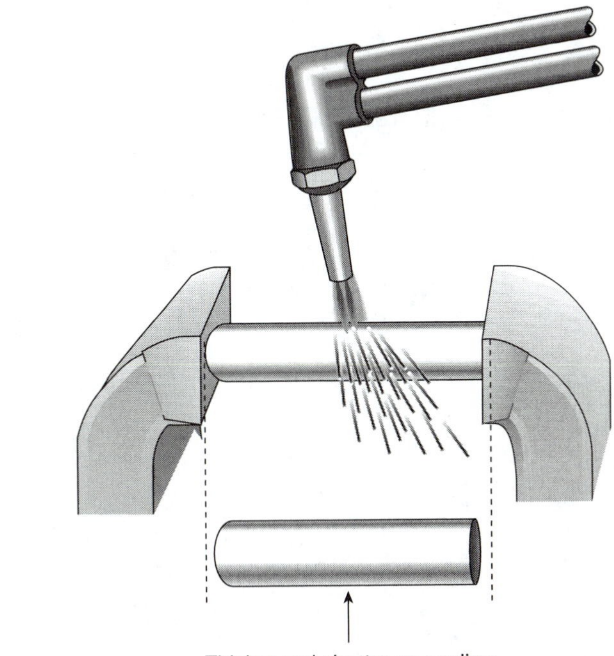

FIGURE 9–2 Restricted heating can affect the size of an object on cooling.

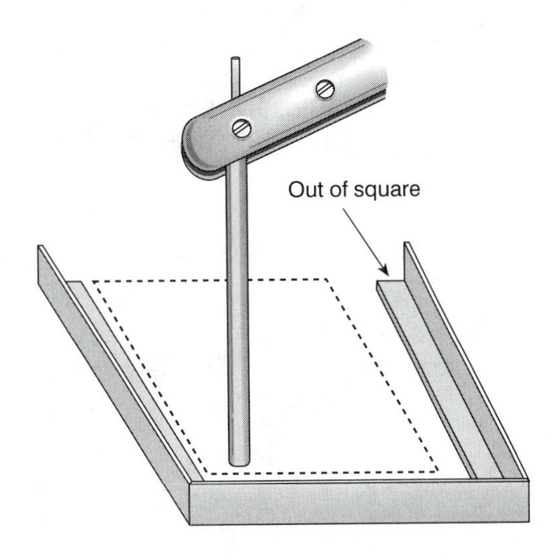

FIGURE 9–3 The effects of distortion caused by failure to use tack welds.

Distortion and the Amount of Welding

The greater the amount of heat generated in the joint, the greater the possibility of distortion. Any time the welder can limit the size of a weld, less heat will result in less distortion. Compare a half-inch fillet weld with a one-inch fillet weld (see Figure 9–4). A one-inch fillet weld has twice the strength of a half-inch fillet weld, but requires four times the filler metal. A welder must consider the added stress that extra filler metal places on the joint. In most applications the thickness of the base metal will determine the size of the weld.

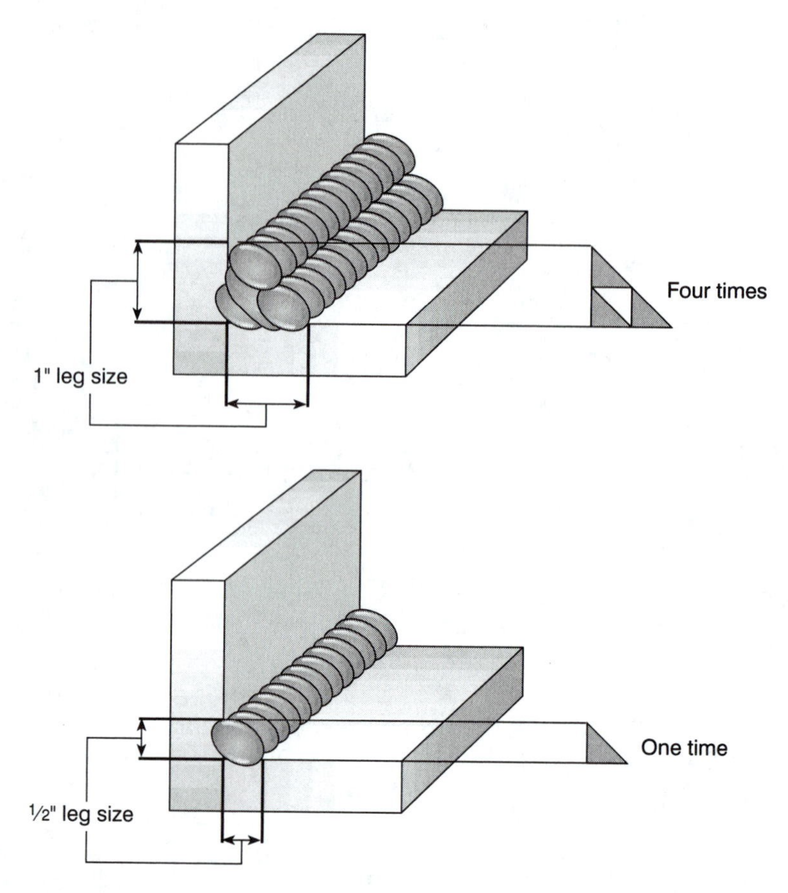

FIGURE 9–4 One-inch weld compared to half-inch weld requires four times more filler metal.

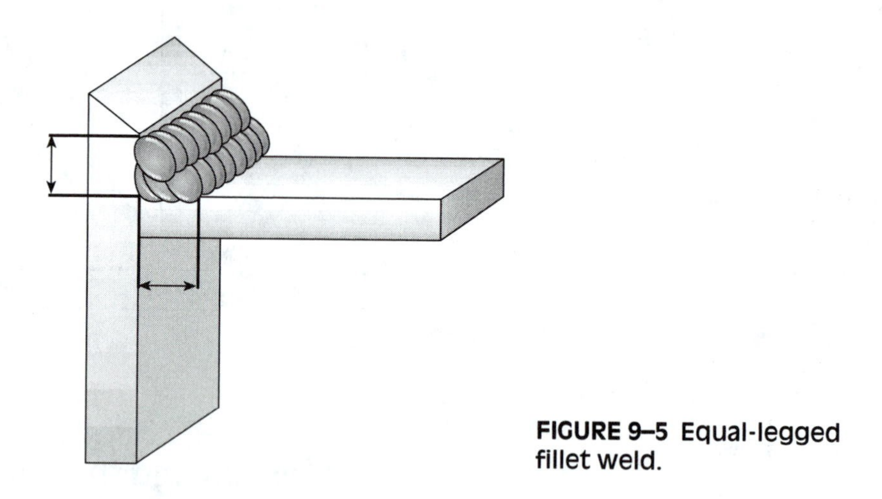

FIGURE 9–5 Equal-legged fillet weld.

Welding produces stress in metal. Yet even metal that has not been welded is stressed through the manufacturing process. Observe this stress by using a hack saw to cut through a small length of tubing or pipe. Before the cut is completed, the blade will become pinched as the cut closes in an effort to relieve some of the stress built into the tubing or pipe during its manufacture. One way to limit stress is to do no more welding than necessary.

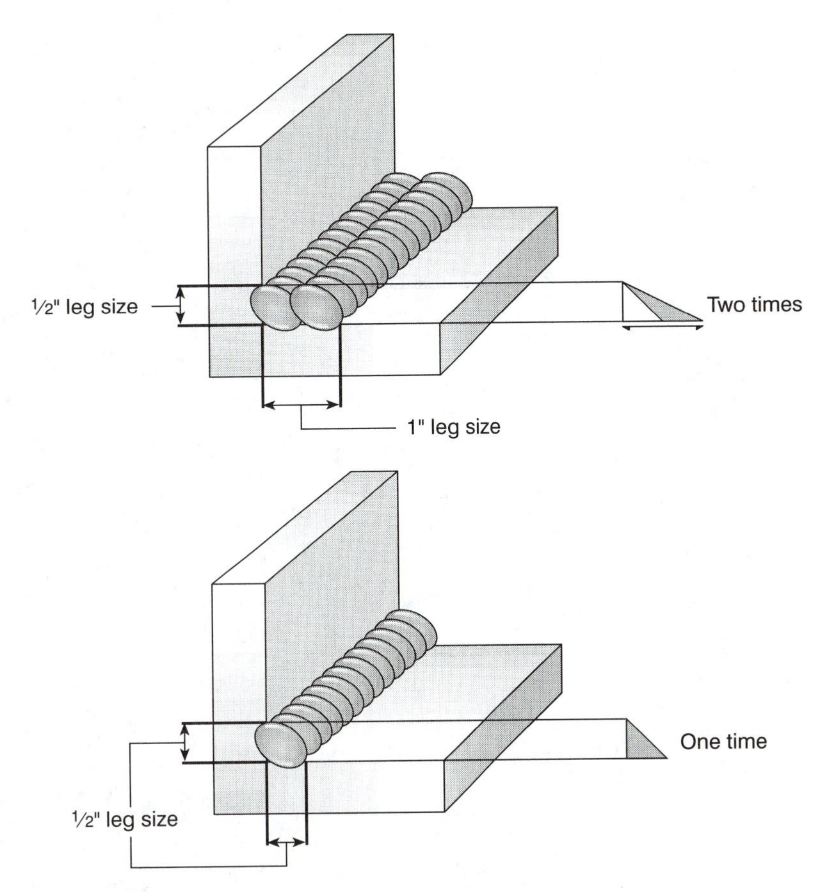

FIGURE 9–6 Unequal-legged half-inch by one-inch fillet weld compared to equal-legged half-inch fillet weld requires twice the filler.

ARC FLASH

While many common stainless steels are nonmagnetic, 400 series stainless steels are magnetic.

Equal leg fillet welds should be used in most applications, as illustrated in Figure 9–5. When the legs are of unequal size, the strength of the weld is decided by the smaller of the two legs. Notice that doubling the size of one leg (½-inch leg by 1-inch leg) will double the filler metal required but add only one-tenth to the **effective throat** (see Figure 9–6). The effective throat is measured as the shortest distance from the weld root to the weld face at the point of a flush contour.

Setup for Welding

A welder must learn how to anticipate the effects welding will have on a given weldment. So be prepared to handle any distortion. Upon cooling, the molten weld pool will contract, pulling on the joint. Keep the following points in mind when welding several pieces together on any fabrication project:

1. Tack weld all of the pieces together (Figure 9–7). Use clamps to hold the weldment in position to keep the tack welds from pulling.

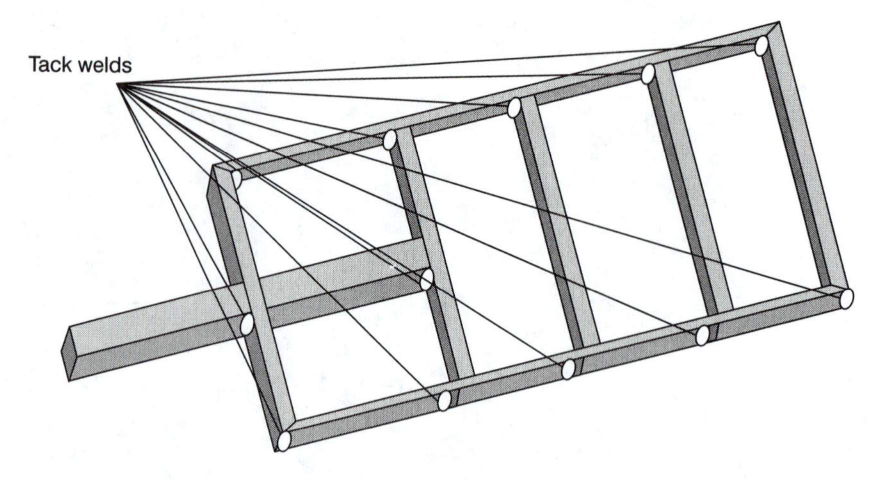

FIGURE 9–7 Trailer assembly tack welded together before being welded.

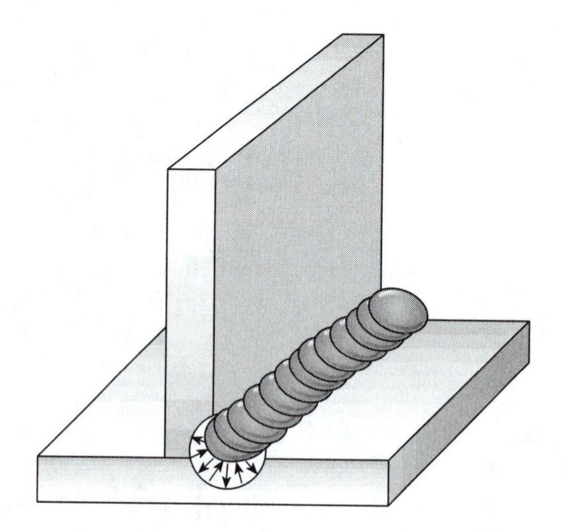

FIGURE 9–8 Welds shrink on cooling.

2. Try to spread out the welding to limit the concentration of heat at any one joint. If the joint is allowed to cool, the effects of heat can be reduced.
3. Do less welding as quickly as possible. Use as large an electrode as possible to reduce the number of passes.
4. Keep the filler metal to a minimum. Do not overweld.
5. While a faster travel speed during welding will reduce heat input, it can cause rapid cooling.
6. Preheating when necessary will provide a more even heat so as to slow cooling and reduce the shrinkage in the welds (Figure 9–8).
7. Cooling between layers in many-layered welds can reduce distortion, but also can change the grain structure of steels sensitive to the formation of martensite.

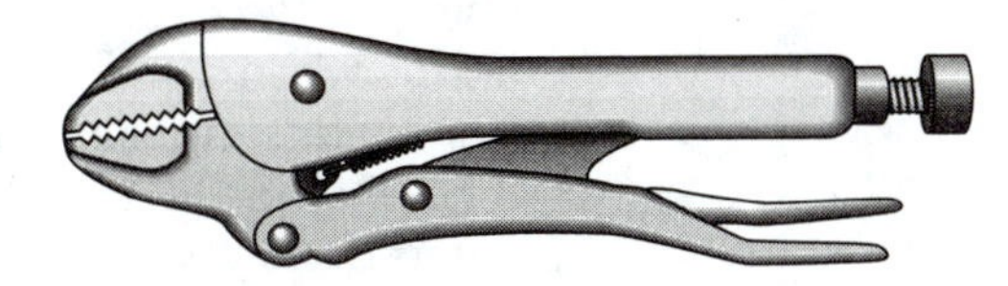

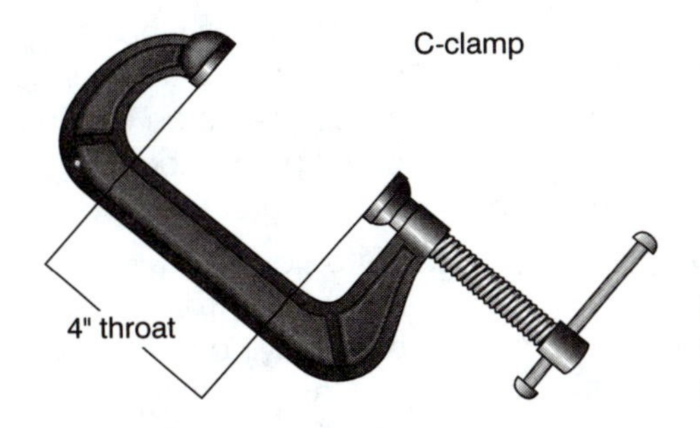

FIGURE 9–9 Handy tools for holding parts in position for welding.

Four 4-inch throated C-clamps and two locking pliers (see Figure 9–9) are the minimum amount of holding tools for any fabrication project. But if these tools are unavailable, don't be afraid to tack weld the project to a plate of steel. Anything that can be used should be used to hold the weldment in position and limit distortion. Remember distortion is a constant companion in welding that never goes away. Learn to limit the effects of distortion caused by welding, and even use distortion to your advantage.

REVIEW

1. Why does distortion always occur as a result of welding?
2. What happens to metal as it is heated? What happens as it cools?
3. What is the purpose of tack welds?

Lab Exercises

1. List the tools in the shop that can be used to control the distortion from welding.
2. Under supervision, use an oxyacetylene flame to straighten a crooked or bent rod.

REVIEW QUESTIONS FOR UNIT 9

Multiple Choice

Choose the best answer.

1. Low-alloy steels:
   a. Are the same as low-carbon steels.
   b. Are often used in mining and agriculture.
   c. Require precautions when welding.
   d. Do not require preheating.
   e. a and c.
   f. b and d.

2. Martensite:
   a. Forms in the HAZ.
   b. Forms when steel cools rapidly.
   c. Can lead to cracking in the joint.
   d. All of the above.
   e. a and b.
3. Metal:
   a. Contracts when heated.
   b. Will return to its original shape if restricted.
   c. Contracts on cooling.
   d. All of the above.
   e. b and c.
4. Distortion:
   a. Results from uneven cooling.
   b. Always occurs in welding to some degree.
   c. Can be limited by tack welds.
   d. All of the above.
   e. b and c.
5. The metal with the highest coefficient of expansion is:
   a. Copper.
   b. Steel.
   c. Aluminum.
   d. Cast iron.
   e. Stainless steel.
6. To limit stress:
   a. Increase heat input.
   b. Use unequal fillet welds in most applications.
   c. Do no more welding than necessary.
   d. All of the above.
   e. None of the above.
7. Steel:
   a. Expands at twice the rate of aluminum.
   b. Is less readily subject to distortion than aluminum.
   c. Is lighter than aluminum.
   d. a and b.
   e. None of the above.
8. Heat:
   a. Contracts metal.
   b. Comes in greater amounts with larger electrodes.
   c. Expands metal.
   d. b and c.
   e. None of the above.
9. The thickness of the base metal:
   a. Has little to do with the amount of welding.
   b. Does not affect heat input.
   c. Usually determines the size of the weld.
   d. All of the above.
   e. None of the above.

Short Answer

1. Does steel or aluminum have a greater coefficient of linear expansion? Does steel or stainless steel?

2. What is the relationship between the size of the weld and distortion?
3. What is stress?
4. How is the size of an unequal leg fillet weld determined?
5. How is the effective throat measured?
6. What happens to the weld pool as it cools?
7. How can tack welds be kept from pulling?
8. How does using larger electrodes affect welding?
9. Why would a ½-inch fillet weld be preferred over a 1-inch fillet weld?
10. What do a C-clamp and locking pliers have in common?

ADDITIONAL ACTIVITIES

Word Find

Circle the technical terms listed below in the letter grid.

```
J M S S T R E S S E A T T
E E T T L O N T N A E N A
T D C A N N O I N A A E C
I I A E C O T R N I D M K
S U R H P A I G D N T D W
N M T T A L K T T L E L E
E C N S Z S A A S D E E L
T A O O D N E N O I P W D
R R C P G H E R E A D P F
A B T A E S T I S Y M A A
M O E R S I A T T E N P C
L N P E X P A N S I O N E
Y O L L A W O L A P X E B
```

CONTRACTS
TACK WELD
STRESS
WELD FACE
WELD ROOT
LOW ALLOY
WELDMENT
MARTENSITE
PREHEAT
POSTHEAT
MEDIUM CARBON
EXPANSION

Writing Skills

1. Write a paragraph describing some of the properties of metal.
2. Write an explanation of what martensite is and what can be done to prevent its formation in welding.
3. Describe what can be done to help prevent distortion.

Math—Decimals

0.1 = one tenth
0.001 = one thousandth
0.00001 = one hundred-thousandth
0.01 = one hundredth
0.0001 = one ten-thousandth
0.000001 = one millionth

In adding decimals, be sure to keep digits in their own proper column. Examples: 0.1 + 0.001 + 0.01 = 0.111   0.0001 + 0.00001 + 0.000001 = 0.000111

1. 0.125 + 2 + 0.45 =

2. 0.025 + 0.0044 + 0.347 + 0.0178 =

3. 0.00051 + 0.0000088 + 0.000009 =

4. 0.0000123 + 0.0000088 + 0.000009 =

In multiplication of decimals, simply multiply the two numbers and count off the total number of decimal places in them. That will be the number in your answer (counting from the right).

Examples: 0.000002 × 5 = 0.000010   0.0000025 × 82 = 0.000205   0.000009 × 20 = 0.00018

5. 0.0000063 × 7 =   **6.** 0.0000123 × 22 =   **7.** 0.000009 × 15 =

The coefficient of linear expansion is a number given to a material to measure its expansion per inch per degree of rise in temperature. Example: If, in the morning a length of aluminum is measured at 16 feet, what is its length at noon with a 53° rise in temperature?

16 × 12 (inches per foot) × 53 × 0.0000123 = 0.1251648, or ⅛ inch. New length 16'⅛".

Coefficient of Linear Expansion:

steel = 0.0000063
aluminum = 0.0000123
stainless steel = 0.000009

8. The steel beam which measured a length of 99 feet in the morning did not fit into place that afternoon. What was its length if its temperature jumped 40 degrees?

9. The stainless steel container, which formerly measured as a 2 foot square, now did not fit into place. Having been heated, its temperature had jumped 290°. What was the amount of increase in its size?

10. The 60-foot aluminum boat had a panel cut for placement in the hull that was 1/16-inch too large and didn't fit. If the panel's temperature was controlled while the temperature of the hull was raised 15°, would that be enough to make the panel fit? Show your work.

Puzzlers

1. A nut had seized on a ½-inch bolt. It was calculated the nut would have to be expanded .003 inch to be removed. How hot would the nut have to be heated and what color would be visible when the nut reached the approximate temperature?

2. Does titanium have a greater coefficient of linear expansion than steel? Explain your answer.

3. An aluminum bar 12 feet long would have to experience a rise of 71° before its length would increase ⅛-inch. How long would a length of steel have to be to experience the same increase of length at 71° (rounded off too the nearest foot)?

UNIT 10

The Welding Project

1. REPAIR AND FABRICATION

Repair

The welding skills required to meet different job opportunities will depend on the amount of training. Training can vary from several hours to several years. Job levels and the pay offered are determined by the skill of the welder. However, whatever the job and the qualification test, the welding will be either repair or fabrication.

So many things are discarded daily that could have been repaired by welding. Welding is one tool that can be used to recycle products that might otherwise be thrown out. Welding can contribute in a positive way to help save what has become a throwaway society. A welder can save energy, save resources, and—when the employer is a large company—save lots of money by doing repairs on things that wear or break. A leak in a pipeline, wear in a shovel bucket, or a break on a piece of earthmoving equipment can often be repaired (Figure 10–1). There is much to be gained by repairing things instead of throwing them away; in some cases there is no alternative.

Not everything repairable by welding can be fixed by the shielded metal arc welding process, though many things can be. Sometimes things do break under stress, wear down, or just rust away. Since the failure of equipment and its parts is an ongoing problem, the repair of metal parts will always require the service of welders. But repairs are not limited to industrial products; home lawn and garden tools can be repaired or redesigned by welding. The cross members of a snowmobile trailer subjected to the corrosion of road salt can be easily replaced by welding without changing the design of the trailer, as in Figure 10–2.

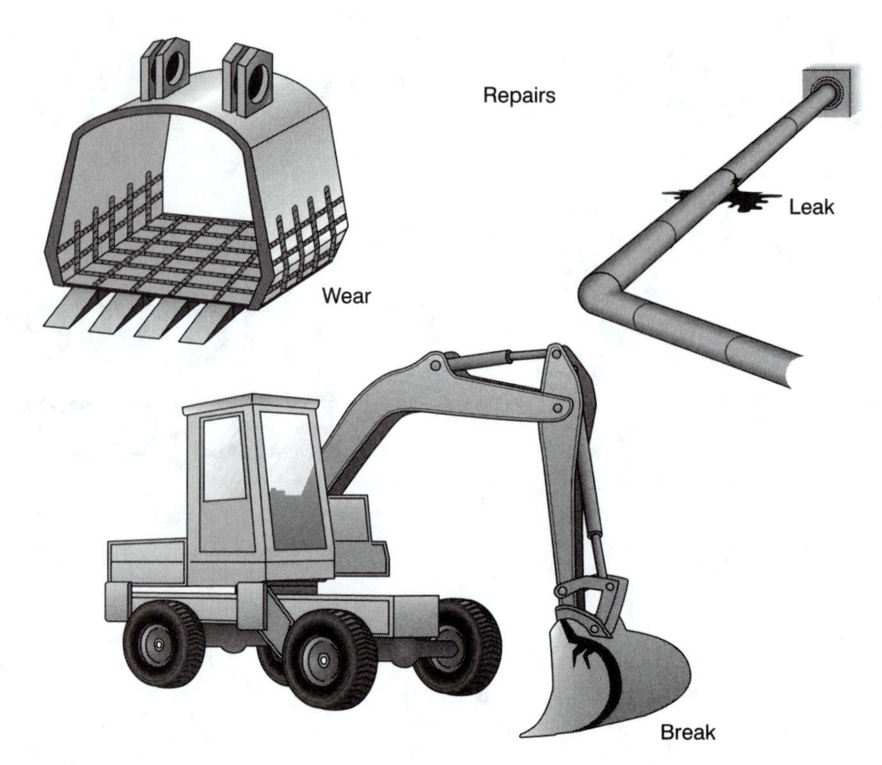

FIGURE 10–1 Repair by welding can take many forms.

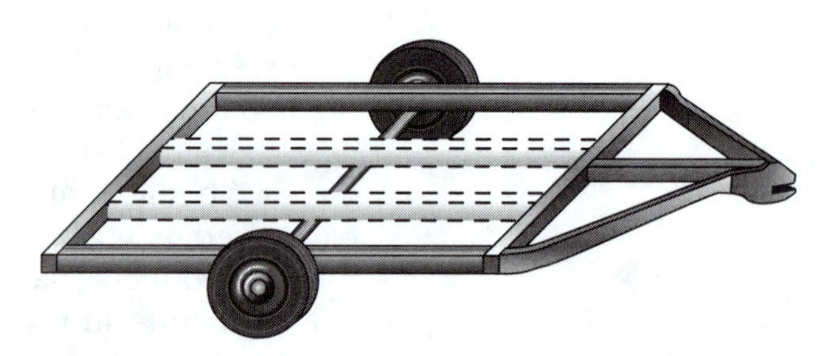

FIGURE 10–2 Trailer rebuilt by welding.

Welding has a major part to play in the repair of costly equipment by extending its serviceable life. There is no need to throw out an otherwise functional tool or piece of machinery that can be repaired by welding.

Fabrication

Fabrication can involve the copying, modification, or creation of entirely new products. Taking an idea and turning it into a useful new product can require welding. Applying professional skills, the welder can transform a few shapes of cold lifeless metal into something that is both strong and practical. Bridges, trucks, and buildings all require the employment of many welders (Figure 10–3).

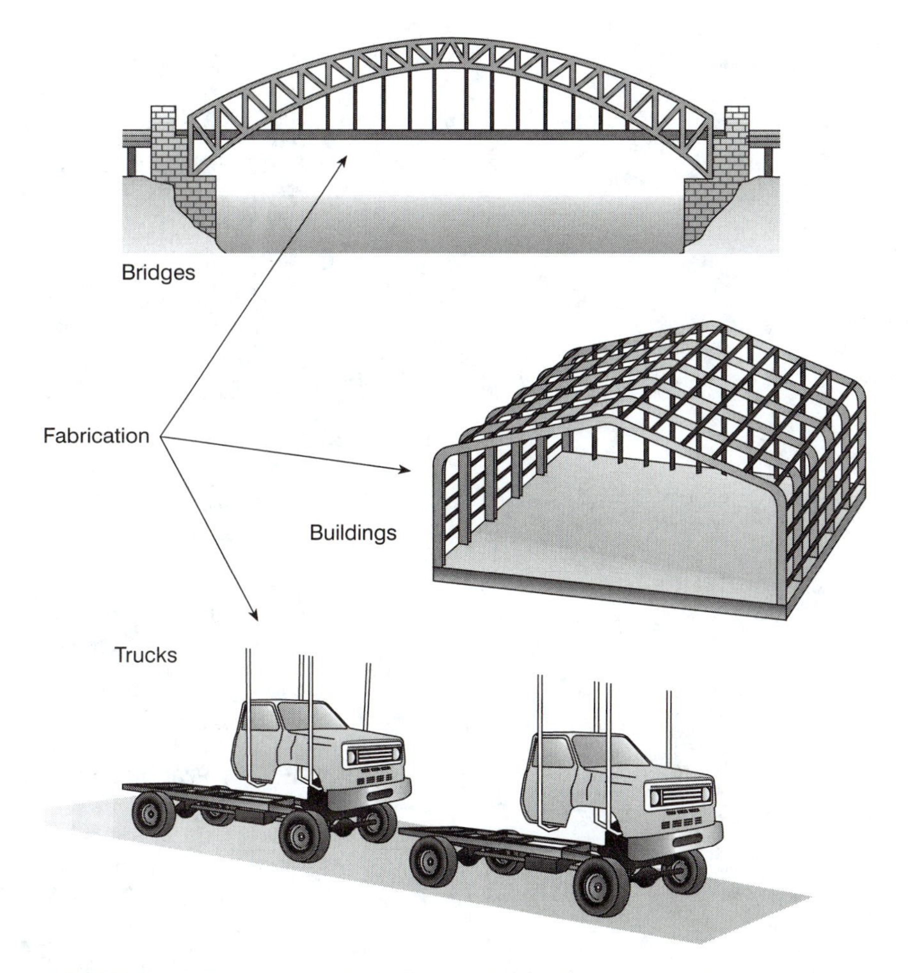

FIGURE 10–3 Commercial fabrication by welding provides many jobs.

In the past, manufacturers constructed overly heavy equipment. Machinery was reinforced with lots of steel. Today's engineering is better at measuring the forces acting on metal. This knowledge has been useful in the manufacturing of products of lighter-weight design. For example, in the automobile industry lighter vehicles produce better gas mileage, but create problems that must be solved to prevent personal injuries of the passengers. Thus, new challenges are created by the use of lighter and more corrosion-resistant metals, with solutions still waiting to be found for the problems these new ideas create. For example, some lighter weight metals are more difficult to weld consistently in high-volume manufacturing processes.

There are thousands of things around the home that can be made using welding equipment. A few of them are pictured in Figure 10–4. A project may require new material, but junked metal from discarded products should not be overlooked. So often the rust is only on the surface and can be removed quite easily by grinding.

Refer to Chapter 1 for the variety of **structural shapes.** A large metal supplier can provide a materials handbook for all the

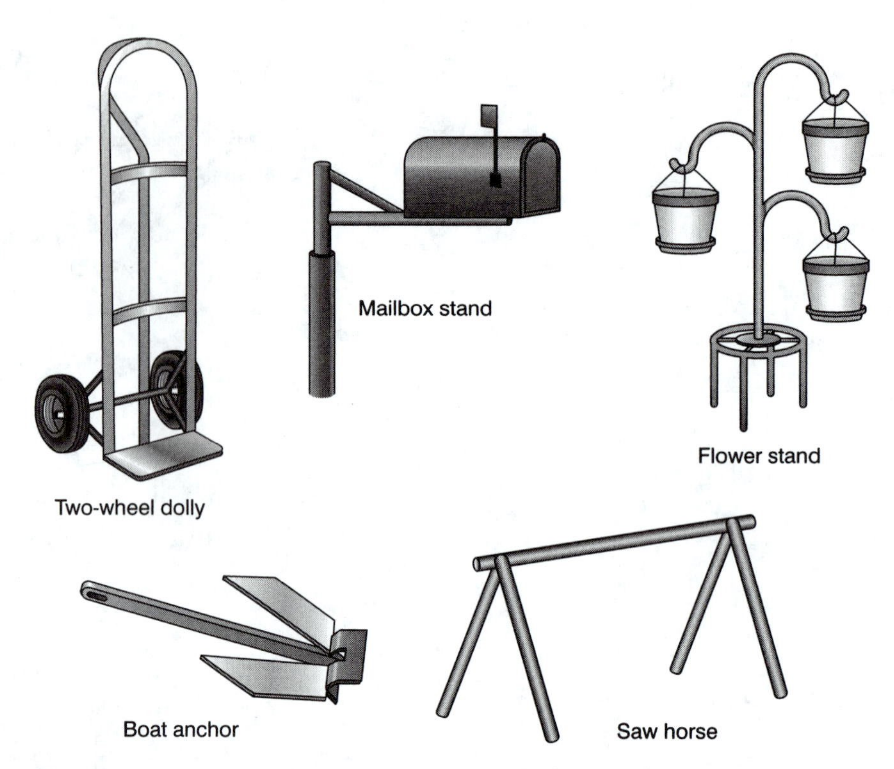

FIGURE 10–4 Practical fabrication projects for the home workshop.

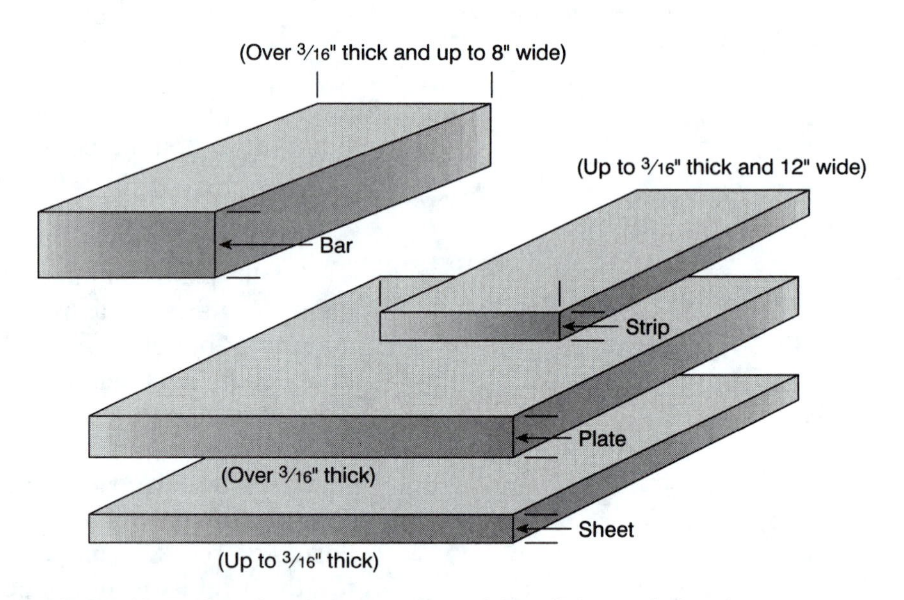

FIGURE 10–5 Structural shapes: bar, strip, plate, and sheet.

shapes they carry. Note how structural shapes are measured (see Figures 10–5, 10–6, and 10–7). When ordering **pipe,** for example, state the *number of lengths* by the *inside diameter* by the *length.* (Pipe up to and including 12 inches is measured by the inside diameter). Buy **hot rolled steel** that is less expensive, not having undergone the extra rolling process that adds to its strength.

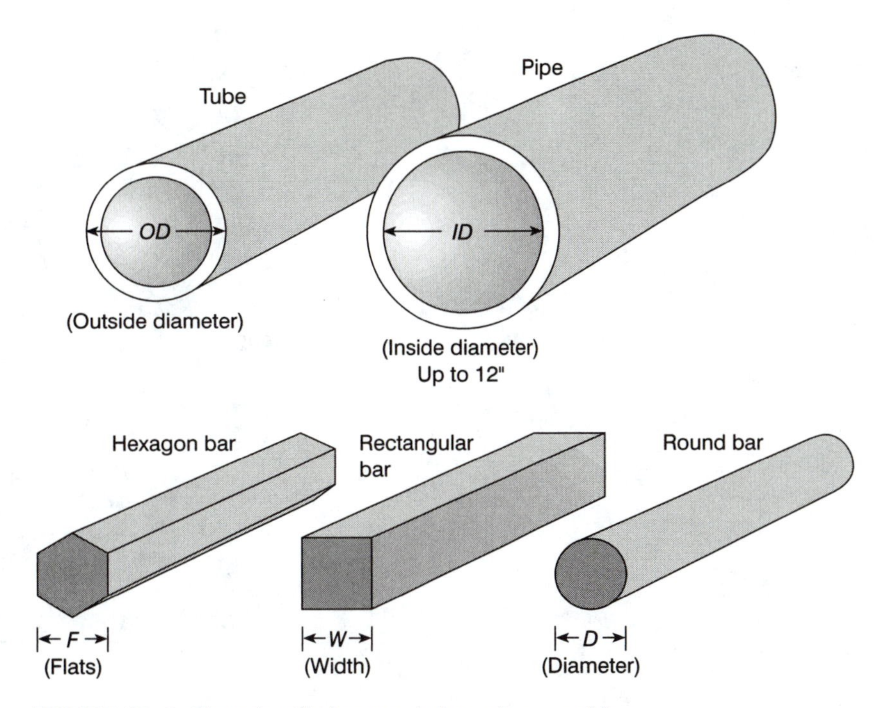

FIGURE 10–6 Structural shapes: tube, pipe, and bar.

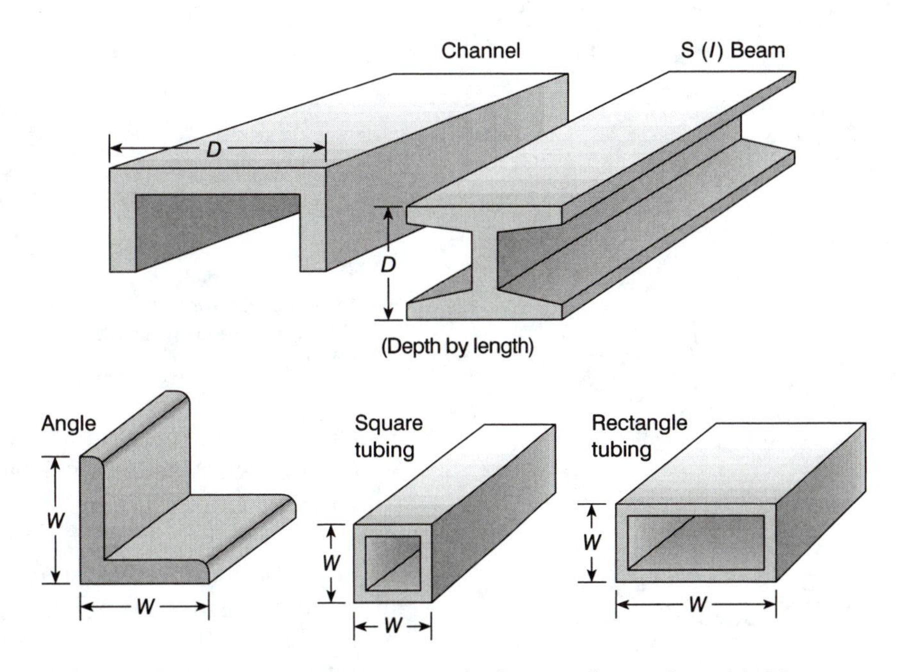

FIGURE 10–7 Structural shapes: angle, beam, channel, and tubing.

Tools

Some tools provide assistance to keep a project in position without moving. Some of the handy tools for holding a weldment in place for welding are the C-clamp, vice grips, and the vice. These tools also prevent the weldment from being pulled out of shape as the welds cool and contract. Because welds will rust quickly when

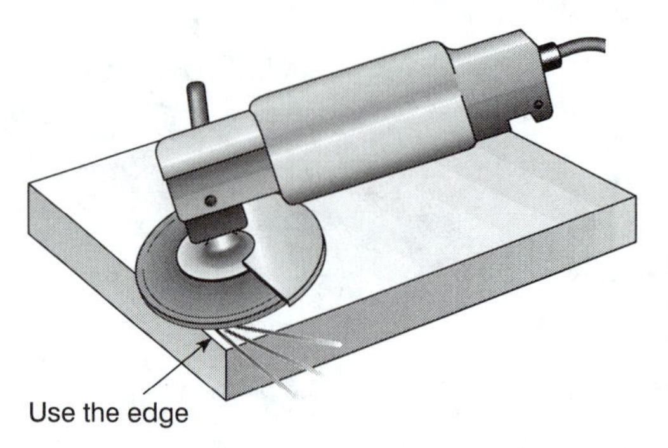

FIGURE 10–8 Proper use of grinder requires contact of the edge on the grinding disk, not the entire bottom surface.

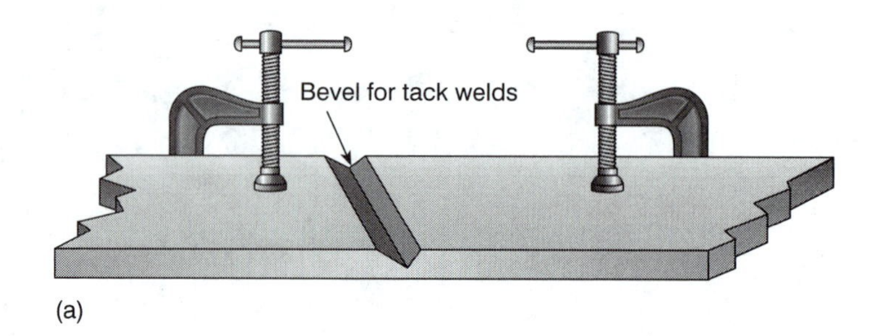

FIGURE 10–9a Position part, clamp, and bevel.
FIGURE 10–9b Tack weld and bevel crack before completing the weld.

exposed to moisture, a coat of primer designed for metal offers protection. A second coat of paint is added over the primer and will give the weldment a finished appearance.

An electric grinder is indispensable in any shop where welding is being done. Rusted metal or metal that has its bluish mill scale covering can be cleaned by grinding or sandblasting before welding. Mill scale, rust, and old welds can be removed quickly by grinding. The grinder also has application for minor jobs that require cutting metal or making a bevel on the edge. The proper technique for using a grinder calls for tilting the grinder so that the bottom of the grinding disk makes contact with the weldment just below the edge, as in Figure 10–8. Do not use the grinder as one would use a polishing tool, where the entire bottom surface of the disk makes contact.

In welding any item where the alignment is critical, tack weld the broken part together and then remove necessary metal in the crack for welding (Figure 10–9). This way when the part is welded

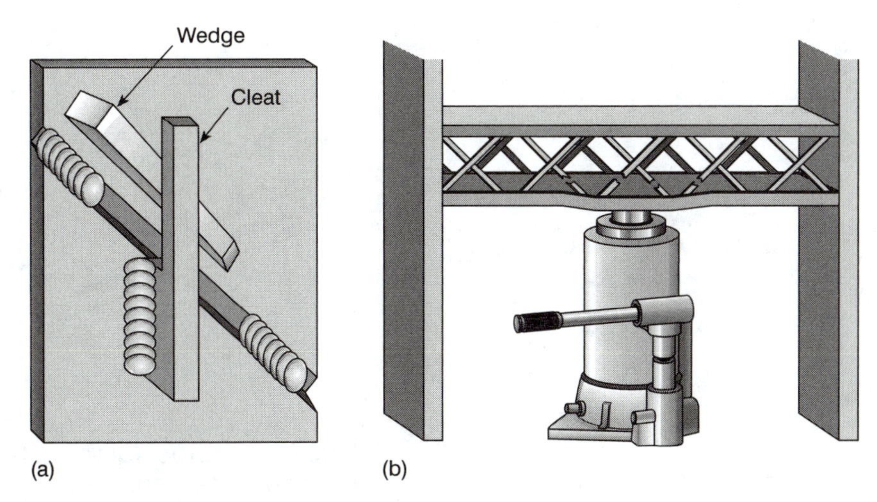

FIGURE 10–10a Use of wedge and cleat for realignment.
FIGURE 10–10b Use of jack for realignment.

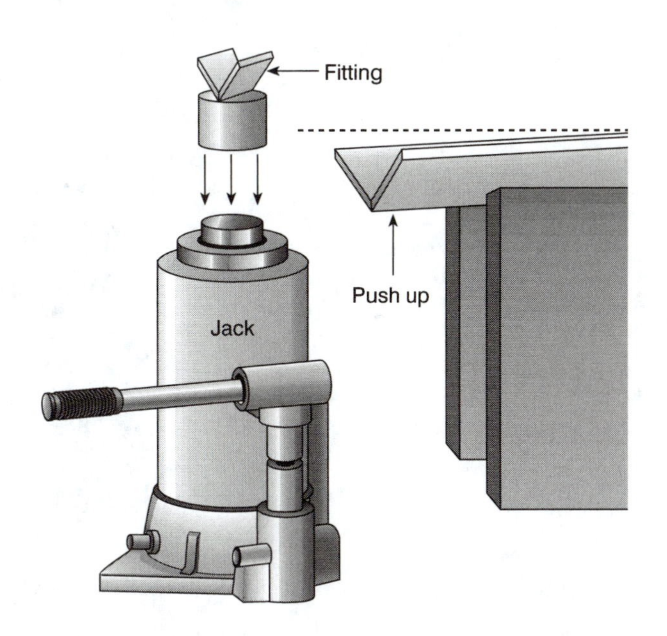

FIGURE 10–11 Use of fabricated fitting for assistance.

back together, it will be aligned. Be careful if making a bevel on a thick part that was broken not to bevel the entire edge, otherwise the part may not fit back together as originally manufactured.

Other tools to be considered for any shop are the jack, and cleats and wedges (illustrated in Figure 10–10). A hydraulic jack may be necessary in repair for bringing parts together. Cleats and wedges can be fabricated for use in bringing parts into alignment. Fittings can be fabricated so a jack will work to apply a controlled force, as in Figure 10–11. Some other tools that are useful include a steel square, soapstone, combination square, and punch, shown in Figure 10–12. Helpful electric tools include the drill, metal cutting saw, and shears (Figure 10–13).

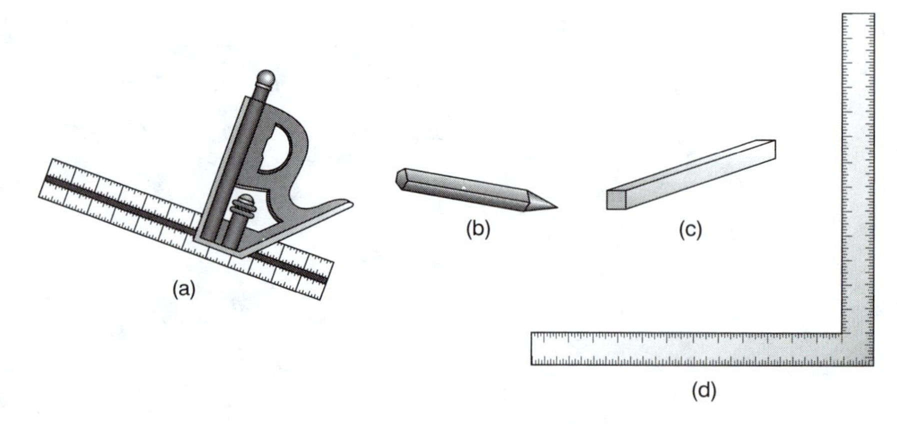

FIGURE 10–12 Useful tools: (a) combination square; (b) punch; (c) soapstone, and (d) steel square.

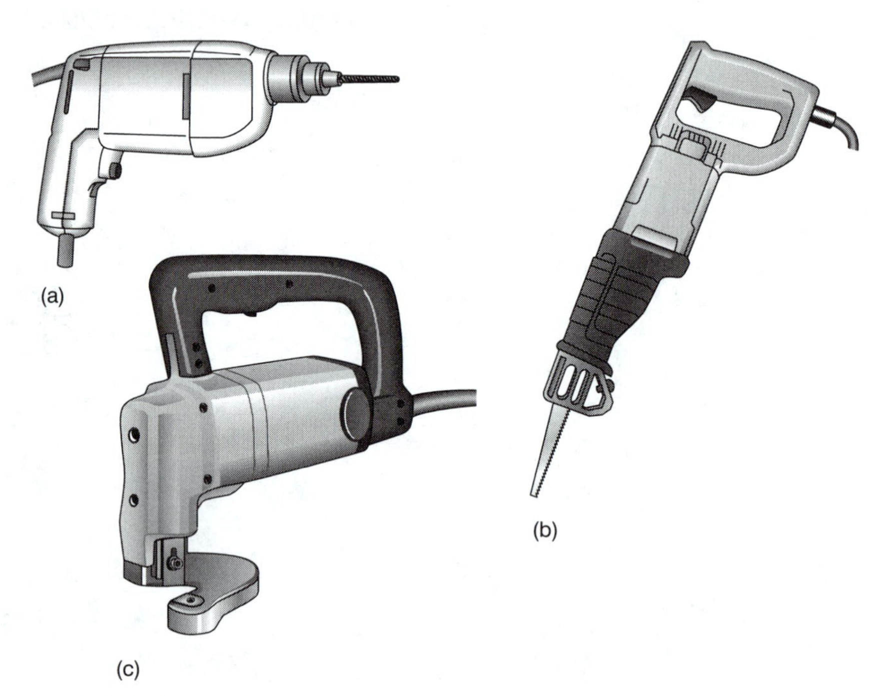

FIGURE 10–13 Useful power tools: (a) electric drill, (b) electric cutting saw, and (c) sheet metal shear.

Layout

When laying out patterns for cutting on sheet or plate, try to work off a straight edge that came from the mill (see Figure 10–14). Check the edge with a framing square or combination square to keep measurements accurate. Then after measuring and marking with soapstone, make your cut on the line or outside the line. Cutting outside the line will be more accurate since grinding the edge will take out any rough spots, bringing the cut back to the measurement as marked.

Finally, after completing the welding, set the weldment aside and come back later with a fresh set of eyes. Are all the required welds in place? If anything is wrong or out of place, a second look often catches the mistake.

ARC FLASH

Make a free-hand sketch of any fabrication project to get a better idea beforehand of what the object is going to look like.

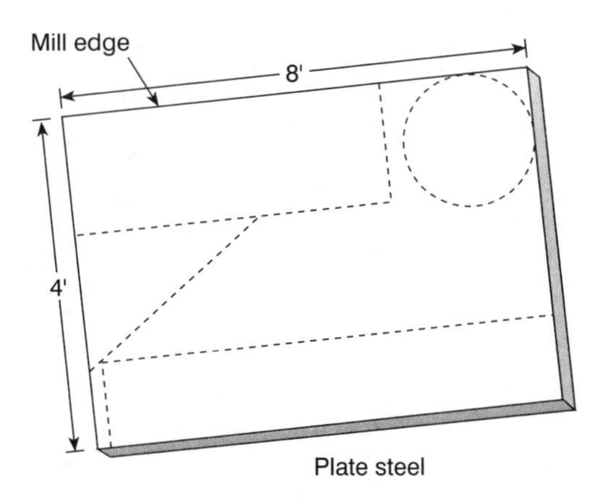

FIGURE 10–14 Organize shapes when cutting, to conserve metal.

SAFETY REMINDER

Be sure paints, thinners, and any other volatile or flammable material are kept out of the shop area.

LAB EXERCISES

1. Repair a project by welding.
2. Sketch out (or computer design) a project you would like to construct by use of welding.
3. Fabricate a project by welding.

APPENDIX A

Advanced Lab Exercises

LAB EXERCISE NUMBER ONE

All of the following exercises can be completed using ¼-inch and ⅜-inch thick steel. The following dimensions are the basic requirements for the exercises: plates ⅜" × 6" × 8" and plates ¼" × 3" × 8". Steel, ⅜-inch thick, is recommended for some exercises, but all the exercises can be done using ⅜-inch material. Begin by cutting up six pieces of 6" × 8" plates and twelve pieces of 3" × 8" plates.

Butt Joint (Designed to Meet AWS Structural Welding Code—Steel D1.1): V-Groove Weld with Backing in Flat Position

The V-groove weld is the popular test given to qualify welders for jobs. Passing this V-groove weld test qualifies applicants for all the other joints in the flat position requiring backing. Passing this test is evidence that you have achieved the skill necessary for moving onto the horizontal position.

Equipment and Material

Safety glasses
Welding gloves
Protective clothing
Amperage: 130–150
Wire brush
Earplugs
E7018 electrodes, ⅛"
Slag hammer
Two pieces ⅜" to 1" A36 steel, 3" × 7" (6" if 1" thick)
One piece ¼" bar, 1" × 8" (backing)
Cutting machine, 22.5° bevel
Two strong backs
Grinder
Helmet

Directions

1. Follow Figure A–1 in preparing the pieces for welding.
2. Grind the surfaces to be welded, including the backing material. Grind the top and bottom ¼-inch beyond the weld area.
3. The fit-up is important before joining pieces together with tack welds. The backing material should have complete contact with the V-groove.
4. The first weld bead requires a slight sideward motion to fuse both pieces together into the backing. Remember to pause momentarily on the sides to avoid undercut.
5. Do not weld over the top edges of the V-groove until you reach the final layer of the weld. Use the edges as a guide for laying the first weld bead of each layer.
6. Air cool. Do not cool in water.
7. Prepare for the guided bend test.

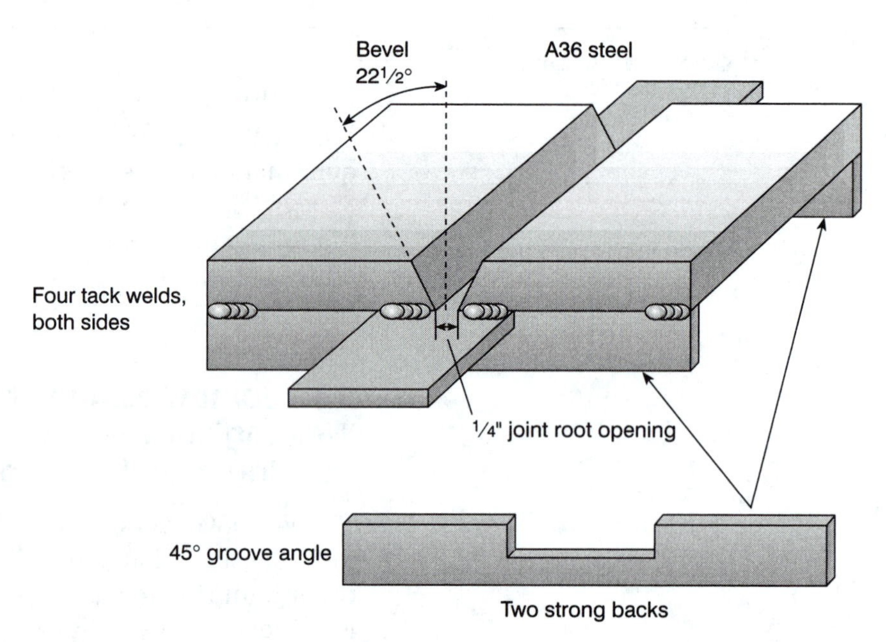

FIGURE A–1 Preparation of V-groove weld with backing in flat position.

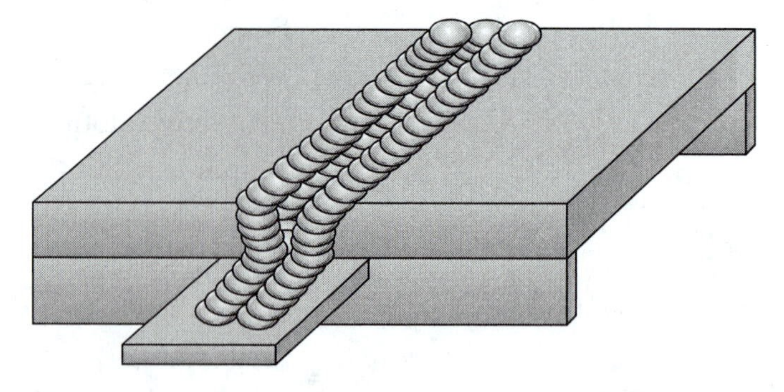

FIGURE A–2 Completed V-groove weld in flat position.

8. First test by visual inspection: check that there is no undercut, underfill, no arc strikes out of the V-groove, appearance has to be of quality (see Figure A–2).
9. Pass the guided bend test. (See Appendix B for details of this test.)
10. Check with the instructor for evaluation, if necessary, before moving on to the next exercise.

LAB EXERCISE NUMBER TWO

Butt Joint (Designed to Meet AWS Structural Welding Code—Steel D1.1): V-Groove Weld with Backing in Horizontal Position

The backing is cut longer to provide an area for beginning (arc strikes) and continuing each weld bead beyond the joint. Arc strikes on the base metal outside of the V-groove can cause weld failure, so watch where you strike the arc. The joint root weld bead of the first pass should be done with a slight sideward motion to fuse both plates to the backing, keeping slag pockets from forming during the joint root pass.

Equipment and Material

Safety glasses
Welding gloves
Protective clothing
Helmet
Slag hammer
Wire brush
Earplugs
E7018 electrode, ⅛"
Amperage: 130–150
Two pieces ⅜" to 1" A36 steel, 3" × 7" (6" if 1" thick)
Cutting machine, 22.5° bevel
One piece ¼" bar, 1" × 8" (backing)
Two strong backs
Grinder

Directions

1. Follow Figure A–3 in preparing the pieces for welding.
2. Grind the surfaces to be welded. Grind the top and bottom ¼ inch beyond the weld area.
3. Fit-up is important before tack welding the pieces together. The backing piece should have complete contact with the V-groove.
4. Do not weld over the top edges of the V-groove until the final cover pass. Use the edges as a guide for making the first weld bead of each layer. Chip off the slag and brush clean after each weld bead.

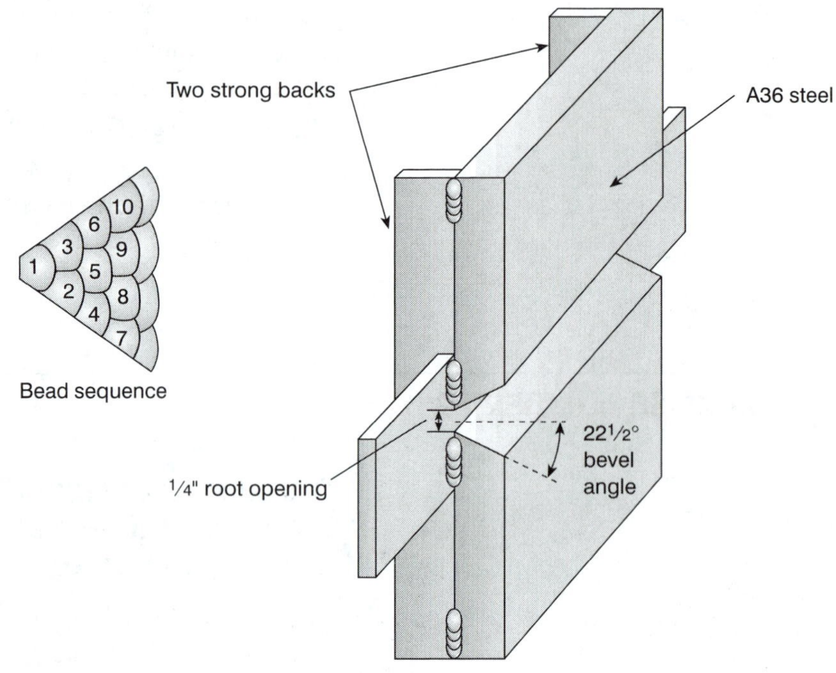

FIGURE A–3 Preparation of V-groove weld with backing in horizontal position.

5. Air cool. Do not cool in water.
6. Prepare for the guided bend test.
7. First test by visual inspection: check for no undercut, underfill, no arc strikes out of the V-groove, appearance has to be of quality.
8. Pass the guided bend test.
9. Check with the instructor for evaluation, if necessary, before moving on to the next exercise.

LAB EXERCISE NUMBER THREE

Padding Plate in Vertical Position—Surfacing Weld

While difficult, the vertical position is the most fun to learn. However, by now your skill is such that welding from the vertical position will be much easier than learning to weld in the flat position was. While there are many different weaving patterns, the zigzag method is easy; it is diagrammed for this exercise. Use uphill welding for the following exercises, as in Figure A–4, unless otherwise indicated. Uphill allows for greater penetration into the base metal than downhill.

Remember to pause momentarily on each side of the joint to avoid undercut. Undercut is the unfilled groove along the weld on each side. A high spot in the middle of the weld may indicate

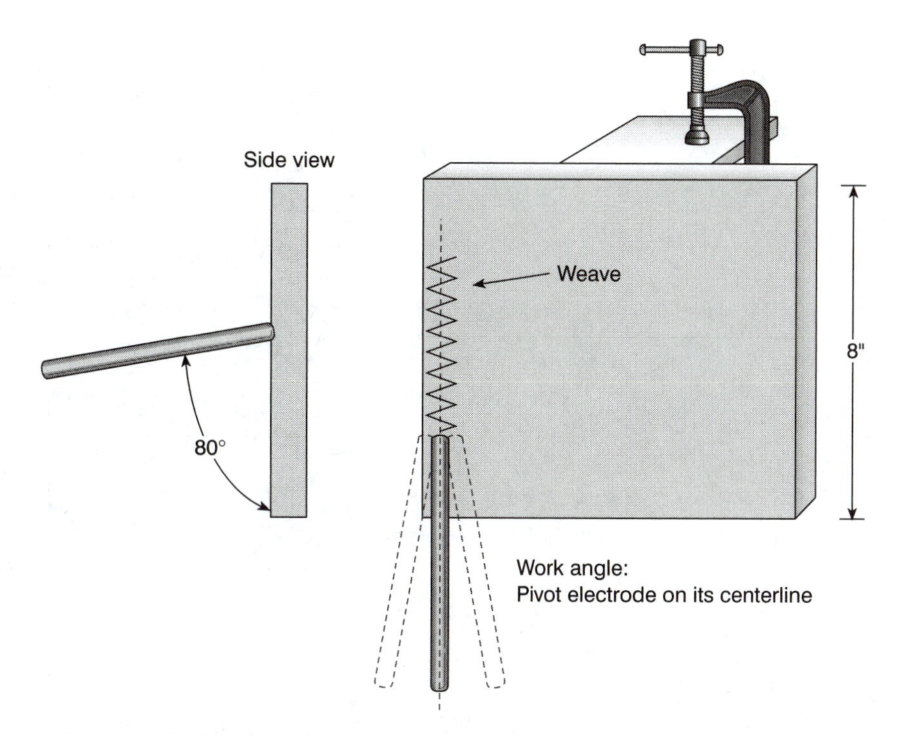

FIGURE A–4 Preparation of padding plate in vertical position.

that the travel speed has been too slow. Speed up across the middle, but always pause on the sides. In practicing this exercise, work on developing a rhythm for the motion of the electrode.

Equipment and Material

Safety glasses
Welding gloves
Protective clothing
Helmet
Slag hammer
Wire brush
Earplugs
Grinder
Amperage:90–110 E6011 electrodes, ⅛"
Amperage: 130–150 E7018 electrodes, ⅛"
One piece ⅜" steel, 6" × 8"

Directions

1. Clean the surface with a grinder.
2. Tack weld a piece of scrap material to the plate. Clamp it in a fixture or weld it to the table at a comfortable height.
3. Weave one layer with E7018 and use a whipping motion for one layer with E6010. When weaving with E7018, the electrode should never leave the weld pool. Do not whip this electrode. In whipping E6010, the electrode comes out of the weld pool and back in again.

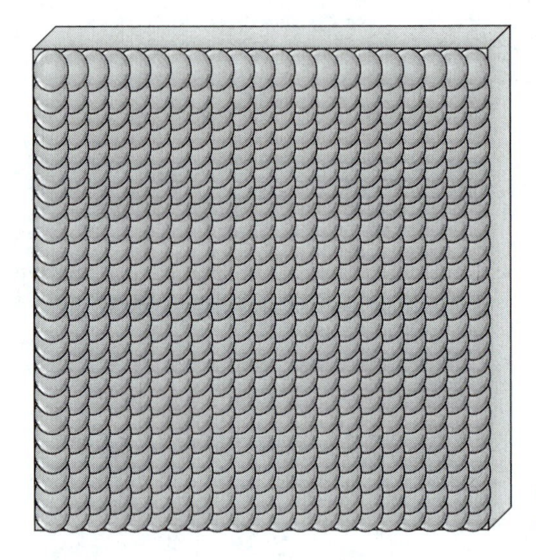

FIGURE A–5 Completed surfacing weld.

4. Begin from the bottom. Practice a weaving method. This is welding uphill.
5. Watch the weld pool. Adjust the travel angle and the work angle when necessary to counteract gravity and maintain control.
6. Avoid the tendency to move too quickly. If necessary, lower the amperage to prevent the molten metal from dripping out of the weld pool.
7. Look for straight weld beads with sufficient lapping of each weld bead. You should find smooth ripples that are neither high in the middle nor undercut on the sides.
8. Cool down the plate when the weld bead begins to drip, or reduce the amperage.
9. Test by visual inspection (see Figure A–5). Practice filling as many padding plates as needed to complete one with the appearance of quality welding.
10. Check with the instructor for evaluation, if necessary, before moving on to the next exercise.

LAB EXERCISE NUMBER FOUR

Tee Joint in Vertical Position—Fillet Weld

This exercise should be relatively easy, after having practiced padding plates using a weaving method. The joint forms a corner (Figure A–6) where the weld pool can flow, readily affected by gravity. Once again remember to pause at each side to avoid undercut. The travel speed across the joint determines the final appearance of a fillet weld. Traveling too slowly causes excessive buildup in the middle.

Equipment and Material

Safety glasses	Earplugs
Welding gloves	Grinder
Protective clothing	E7018 electrode, ⅛"
Helmet	Amperage:130–150
Slag hammer	Two pieces ¼" to ⅜" steel, 6" × 3"
Wire brush	

Directions

1. Clean the surfaces with a grinder.
2. Tack weld two pieces together at each end.
3. Clamp the joint in a fixture or tack weld to the table at a comfortable height.
4. Beginning at the bottom, begin practicing a weave on the second pass. The welding will be made uphill.
5. Avoid the tendency to move too fast. Use the proper techniques (electrode angles and amperage setting) to keep the molten metal from dripping.
6. Make four or five layers on each side. Cool down the preceding layer so the weld pool is easier to control; remember the joint is not large enough to disperse the heat properly. Chip off the slag and brush clean after making each weld bead.
7. Look for consistent ripples with no gaps, a smooth weld face, and no undercut.

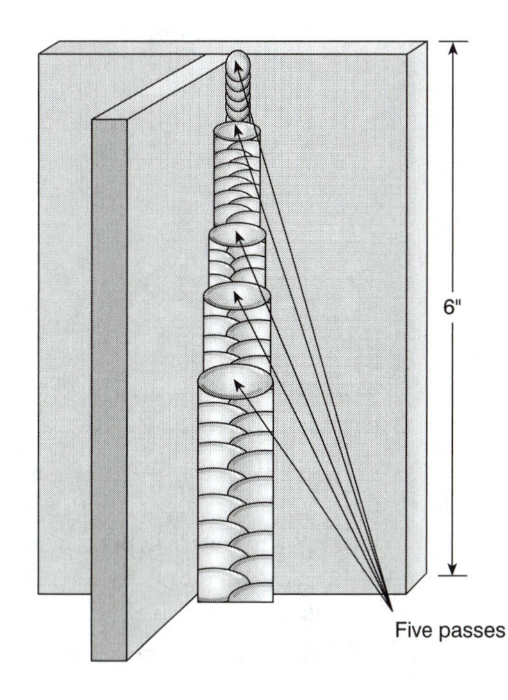

FIGURE A–6 Tee joint in vertical position.

8. Test by visual inspection. Practice this exercise until you are confident that the appearance of your fillet weld is of quality.
9. Check with the instructor for evaluation, if necessary, before moving on to the next exercise.

LAB EXERCISE NUMBER FIVE

Butt Joint (Designed to Meet AWS Structural Welding Code—Steel D1.1): V-Groove Weld with Backing in Vertical Position

Begin the arc below the joint on the backing material. Avoid arc strikes made on the surface of the joint outside the weld. Arc strikes outside the weld should cause the weldment to be rejected upon visual examination.

Take care to clean weld bead completely when making restarts, or arc strikes in the testing area. Trapped slag can cause weld failure. The joint root weld bead is so very important; be sure there is complete fusion along the entire length of weld.

Equipment and Material

Safety glasses	Amperage: 130–150
Welding gloves	One piece ¼" bar, 1" × 8" (backing)
Protective clothing	Two strong backs
Two pieces ⅛" to 1" A36 steel, 3" × 7" (6" if 1" thick)	Grinder
Wire brush	Cutting machine, 22.5° bevel
Earplugs	Helmet
E7018 electrode, ⅛"	Slag hammer

Directions

1. Follow Figure A–7 in preparing the pieces for welding.
2. Grind the surfaces to be welded. Grind the top and bottom of each plate ¼ inch beyond the weld area.
3. Fit-up is important. The backing piece should have complete contact with the V-groove before you make the tack welds.
4. Begin at the bottom. All weld beads will be completed uphill.
5. Do not weld over the top edges of the V-groove until the final cover pass. Use the edges as a guide in laying the first weld bead of each layer. Chip away the slag and brush clean after each weld bead.
6. Air cool. Do not cool in water.

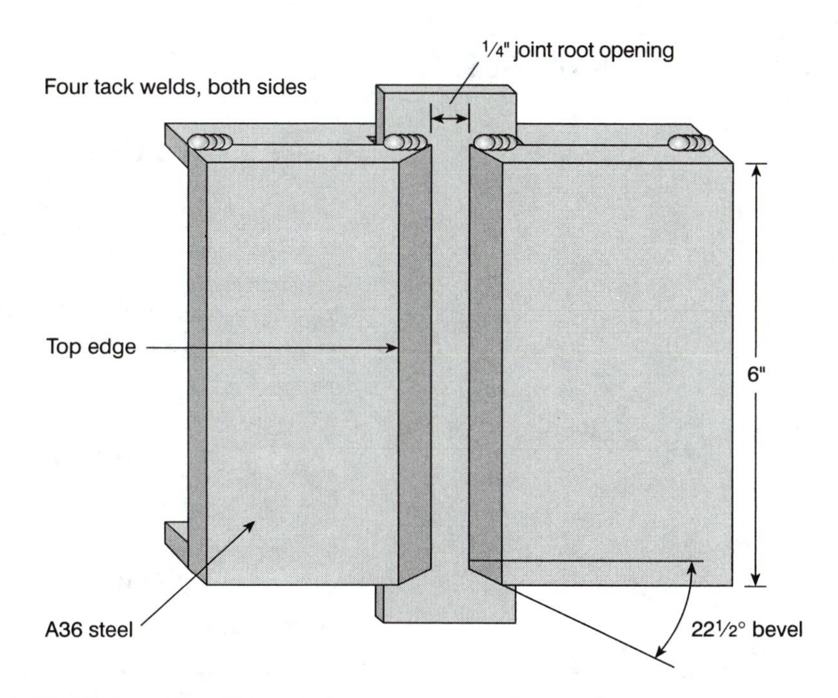

FIGURE A–7 Preparation of V-groove weld with backing in vertical position.

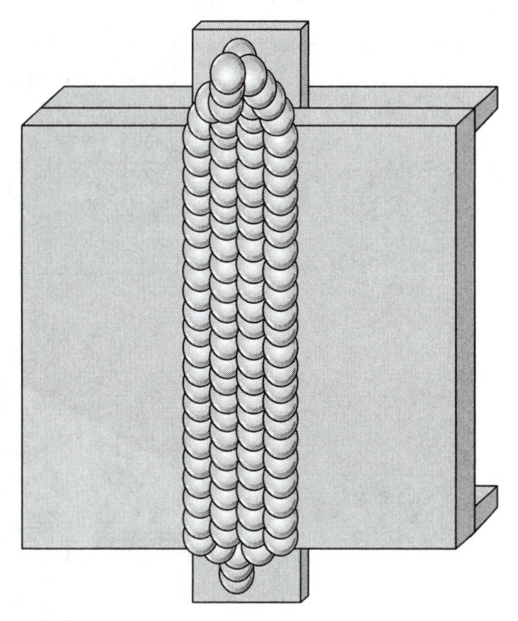

FIGURE A–8 Completed V-groove weld in vertical position.

7. First test by visual inspection: no undercut or underfill, no arc strikes out of the V-groove, appearance has to be of quality, as in Figure A–8.
8. Pass the guided bend test.
9. Check with the instructor for evaluation, if necessary, before moving on to the next exercise.

Tee Joint in Overhead Position—Fillet Weld

The joint root weld bead will involve a slight sideward motion to fuse both pieces. All other weld beads must be stringers. Find a comfortable position from which to view the weld pool. Position your helmet right under the work. Wearing leathers or extra protection on the upper body is recommended when welding in the overhead position (see Figure A–9).

Equipment and Material

Safety glasses
Welding gloves
Protective clothing
Slag hammer
Wire brush
Earplugs
Grinder
E7018 electrode, ⅛"
Amperage: 130–150
Two pieces ¼" to ⅜" steel, 6" × 3"
Fixture for positioning
Helmet

Directions

1. Clean the surfaces.
2. Tack weld two pieces together at each end.
3. Tack weld a piece of scrap material to the joint, if needed. Clamp it in a fixture.
4. Begin on one side. Watch the weld pool and maintain a 10° travel angle.

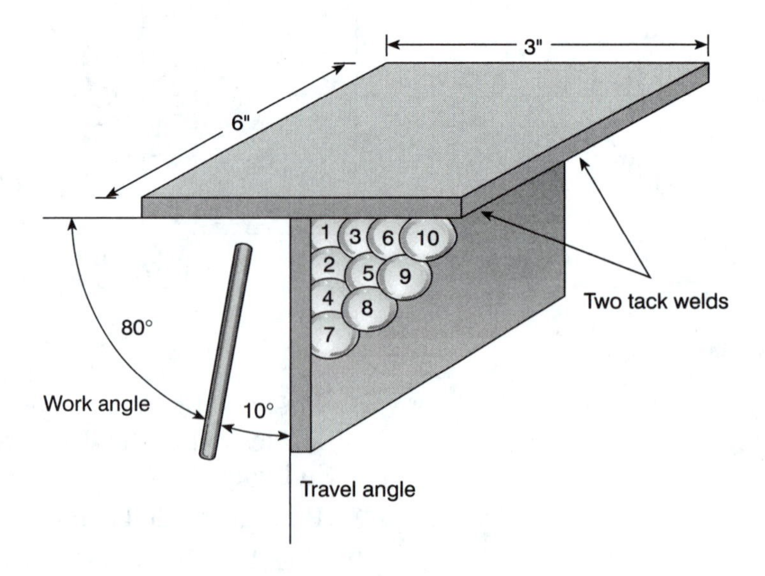

FIGURE A–9 Tee joint in overhead position.

5. Avoid the tendency to move too quickly. Use of the proper techniques (angles and amperage setting) will keep the weld from dripping.
6. Make four or five layers on each side.
7. Look for even weld beads with consistent ripples, lapping weld beads, and no undercut.
8. First test by visual inspection: no undercut, underfill, no arc strikes out of the V-groove, appearance has to be of quality.
9. Practice this exercise until you meet the standards established for this course.
10. Check with the instructor for evaluation, if necessary, before moving on to the next exercise.

LAB EXERCISE NUMBER SEVEN

Butt Joint (Designed to Meet AWS Structural Welding Code—Steel D1.1): V-Groove Weld with Backing in Overhead Position

Do not hurry the travel speed especially on the joint root weld bead. Wash (touch) the edges of each piece with the weld pool on the joint root pass. While this first pass is completed with a slight sideward motion, all the other weld beads should be stringers. Watch the work angle, the travel angle, and the arc length. Too great a travel angle will reduce root penetration and produce a weld face that is high in the middle. Too long an arc length will cause unnecessary spatter. Leathers or extra protection on the upper body is recommended for this exercise (Figure A–10).

Equipment and Material

Safety glasses
Welding gloves
Helmet
Slag hammer
Wire brush
Earplugs
E7018 electrode, ⅛"
Protective clothing

Two pieces ⅜" to 1" A36 steel, 3" × 7" (6" if 1" thick)
Grinder
Amperage: 130–150
Cutting machine, 22.5° bevel
One piece ¼" bar, 1" × 8" (backing)

Directions

1. Follow Figure A–10 in preparing the pieces for welding.
2. Grind the surfaces to be welded. Grind the top and bottom ¼ inch beyond the weld area.

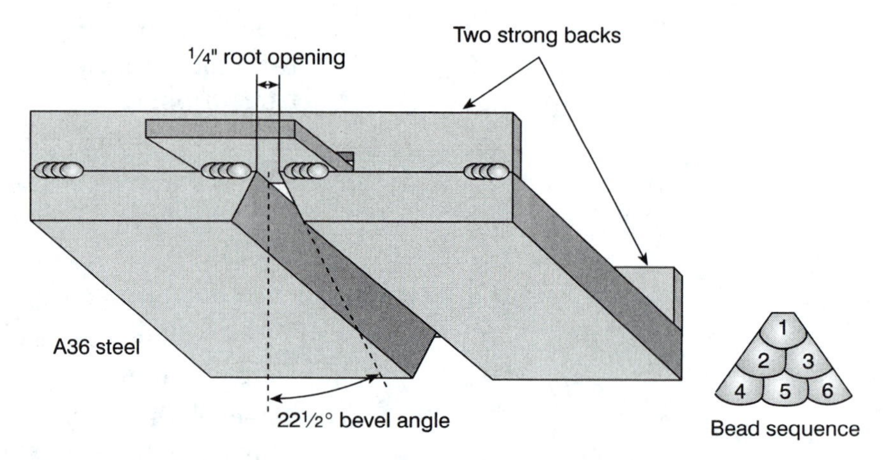

FIGURE A–10 Preparation of V-groove weld with backing in overhead position.

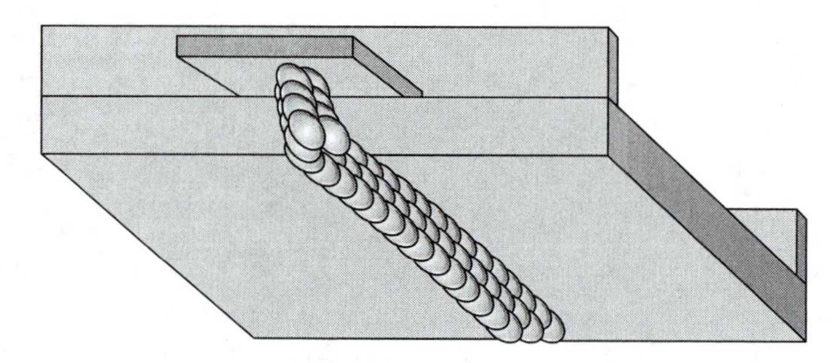

FIGURE A–11 Completed V-groove weld in overhead position.

3. Fit-up is important before making tack welds. The backing piece should have complete contact with the V-groove.
4. Do not weld over the top edges of the V-groove until the final cover pass. Use the edges as a guide for making the first weld bead of each layer.
5. Chip off the slag and brush clean after each weld bead.
6. Air cool. Do not cool in water.
7. First test by visual inspection: no undercut, no underfill, no arc strikes out of the V-groove, appearance has to be of quality, as in Figure A–11.
8. Pass the guided bend test.
9. Check with the instructor for evaluation, if necessary, before moving on to the next exercise.

Butt Joint (Designed to Meet AWS Structural Welding Code—Steel D1.1): V-Groove Weld without Backing

The V-groove weld without backing will help prepare you for pipe welding. The setup is the same for doing all these exercises without backing in the flat position, horizontal position, vertical position and overhead position.

The keyhole (a partial melting of the edges on both pieces) is the basis for successfully welding a V-groove weld without backing. The keyhole is necessary for achieving complete root penetration. Use a whipping motion for the E6010 electrode. Once again, whipping is a quick motion out of the weld pool and back in again, pushing the keyhole along the joint root (Figure A–12).

Equipment and Material

Safety glasses

Protective clothing

Helmet

Slag hammer

Grinder

E6010 electrode, ⅛"

Earplugs

Wire brush

Two pieces ⅜" to 1" A36 steel, 3" × 7" (6" if 1" thick)

Cutting machine, 30° bevel

Welding gloves

Two strong backs

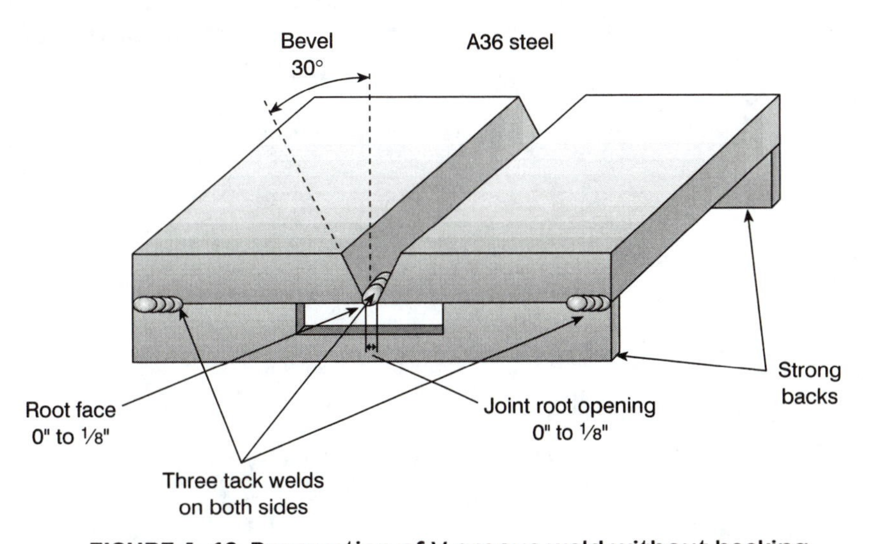

FIGURE A–12 Preparation of V-groove weld without backing.

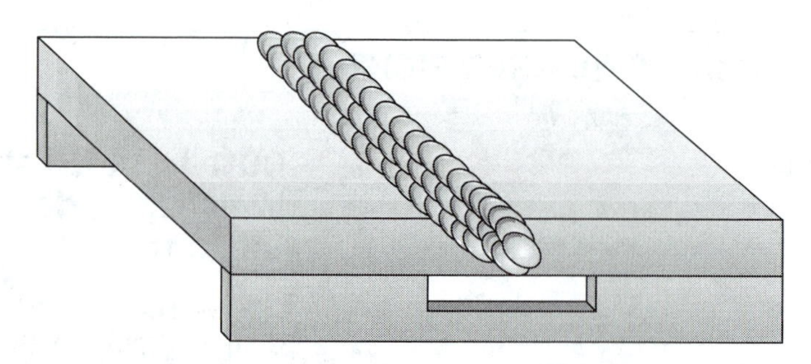

FIGURE A–13 Completed V-groove weld without backing.

Directions

1. Follow Figure A–12 in preparing the pieces in each of the four welding positions.
2. Grind the surfaces to be welded. Grind the top and bottom ¼ inch beyond the weld area.
3. Fit up, align, and tack weld.
4. The first bead (joint root pass) is critical. The success of this joint depends on the joint root pass. Use a whipping motion with E6010 (this is not a weaving motion).
5. Do not weld over the top edges of the V-groove until the final cover pass. Use the edges as a guide for laying the first weld bead of each layer. Always chip away the slag and brush clean between each weld bead.
6. After completing one side, turn the weld over and back gouge out (remove) the root bead before finishing with a back weld.
7. First test by visual inspection: no undercut, underfill, no arc strikes out of the V-groove, appearance has to be of quality, as in Figure A–13.
8. Pass the guided bend test.
9. Check with the instructor for evaluation, if necessary, before moving on to the next exercise.

APPENDIX B

Guided Bend Test

Introduction

The **guided bend test** is a method for examining a welder's ability to follow a given procedure. This is a destructive test in which coupons are bent, stretching them to examine for flaws in the welding. The guided bend test is used for making face bends, root bends, or side bends. The **face bend** stretches the weld face. The **root bend** stretches the root. The **side bend** stretches the side. These are shown in Figure B–1.

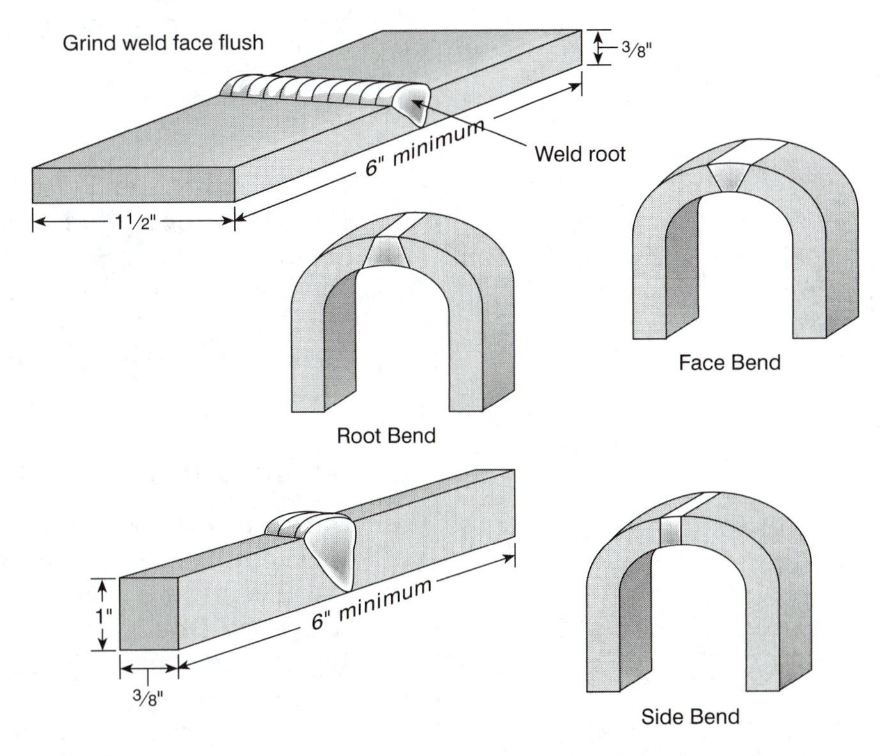

FIGURE B–1 Preparation for root bend, face bend, and side bend.

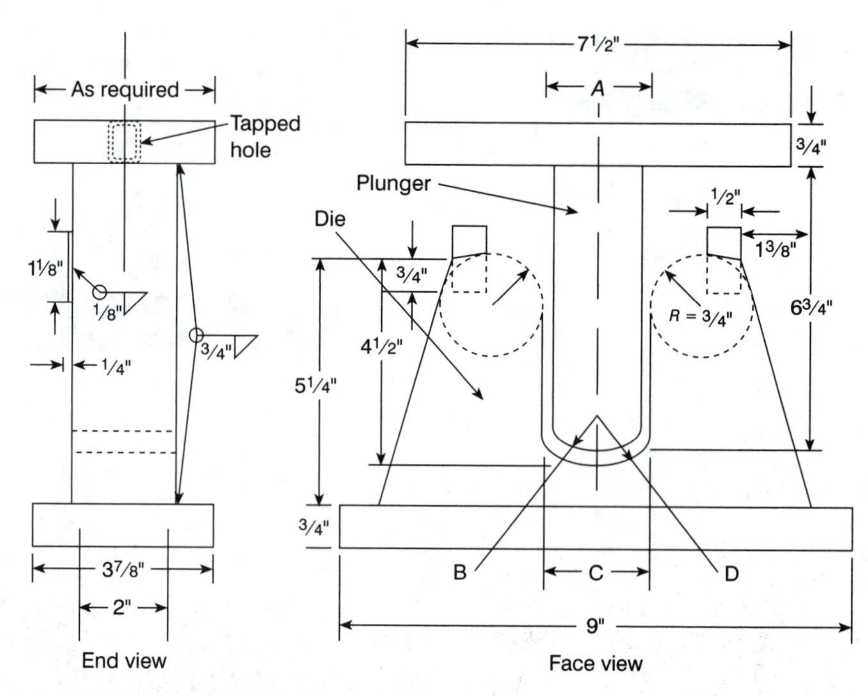

FIGURE B–2 Jig design.

TABLE B–1 Dimensions for three jigs.

Yield Strength (psi)	*A*	*B*	*C*	*D*
50,000 and under	1 ½"	¾"	2 ⅜"	1 ³⁄₁₆"
50,000 to 90,000	2"	1"	2 ⅞"	1 ⁷⁄₁₆"
Greater than 90,000	2 ½"	1 ¼"	3 ⅜"	1 ¹¹⁄₁₆"

The guided bend test requires a jig for bending the coupons. The design for a jig illustrated in Figure B–2 is specified by the American Welding Society's *Structural Welding Code Steel D1.1*. Those specs are given in Table B–1. A hydraulic jack is used to push either the shoulder or the plunger onto the coupon, bending and stretching the material. A36 steel, if it is the type being bent, has a yield strength of 36,000 psi minimum.

Qualifications of Testing

The specifications are given for testing groove welds of limited thickness, in Figure B–3, and of unlimited thickness, in Figure B–4. The groove weld test of the butt joint, according to the *Structural Welding Code Steel D1.1*, can be given in any of the four positions for welding, with the limitations dependent upon the thickness of the base metal (see Table B–2). Successful completion of tests in the vertical and overhead positions qualifies a welder in all four positions.

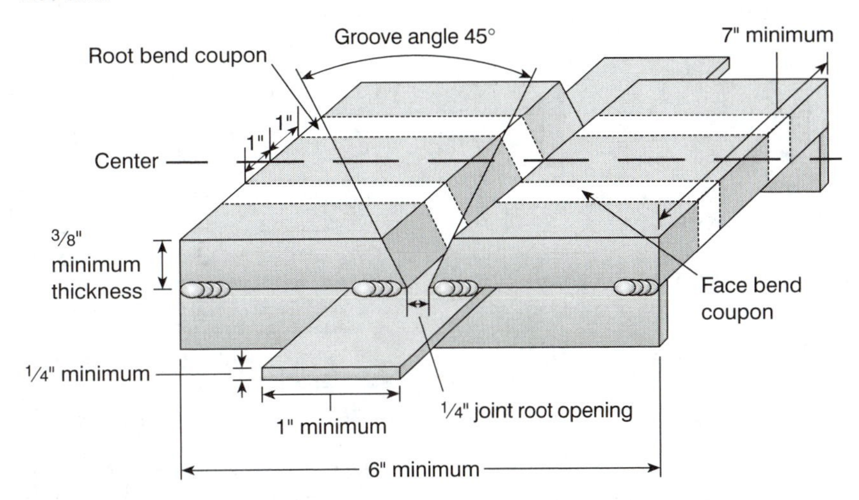

FIGURE B–3 Specification for groove welds of limited thickness.

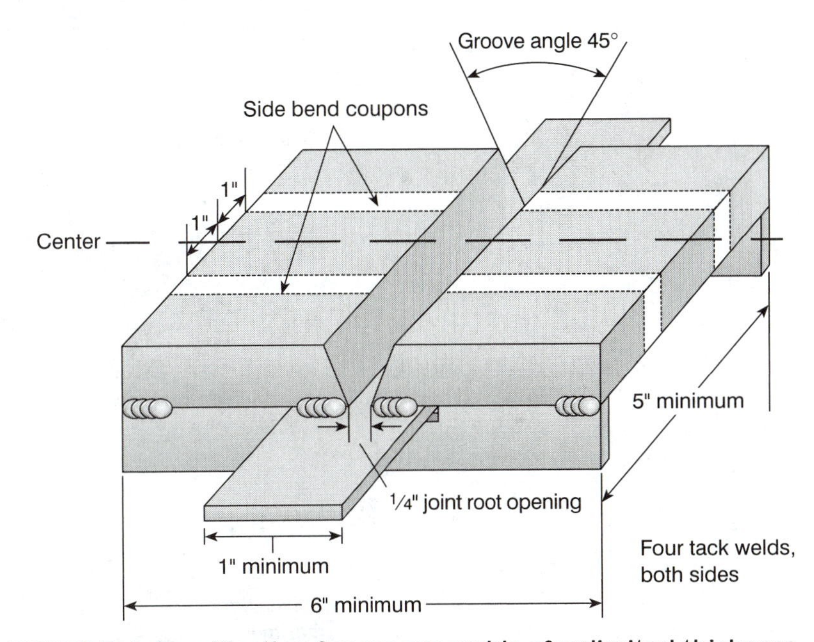

FIGURE B–4 Specification for groove welds of unlimited thickness.

Preparing Coupons for Testing

The coupons must be taken from designated areas in the weldment as noted in the specifications given in Figures B–3 and B–4. A jig designed for 50,000 psi and under meets the requirements of A36 steel with a yield strength of 36,000 psi minimum.

In preparing coupons for testing, they can be cut to size with a torch or a band saw. Reinforcement on the weld face, root face, backing materials, and strong backs must be removed; this can be done with a torch. Grind the weld flush using a hand grinder or a pedestal grinder, leaving the grinding marks parallel to the length of the coupon. File the edges of the coupon to a ⅛-inch maximum radius, as shown in Figure B–5.

TABLE B–2 Welding limits for successful testing of ⅜-inch groove welds.

Position Tested	*Type of Weld*	*Position Qualified*	*Thickness Qualified*
Flat	Groove Weld	flat	Up to ¾"
Horizontal	Groove Weld	flat and horizontal	Up to ¾"
Vertical	Groove Weld	flat, horizontal, and vertical	Up to ¾"
Overhead	Groove Weld	flat and overhead	Up to ¾"
Welding Limits for Successful Testing of 1-inch Groove Welds			
Flat	Groove Weld	flat	Unlimited
Horizontal	Groove Weld	flat and horizontal	Unlimited
Vertical	Groove Weld	flat, horizontal, and vertical	Unlimited
Overhead	Groove Weld	flat and overhead	Unlimited

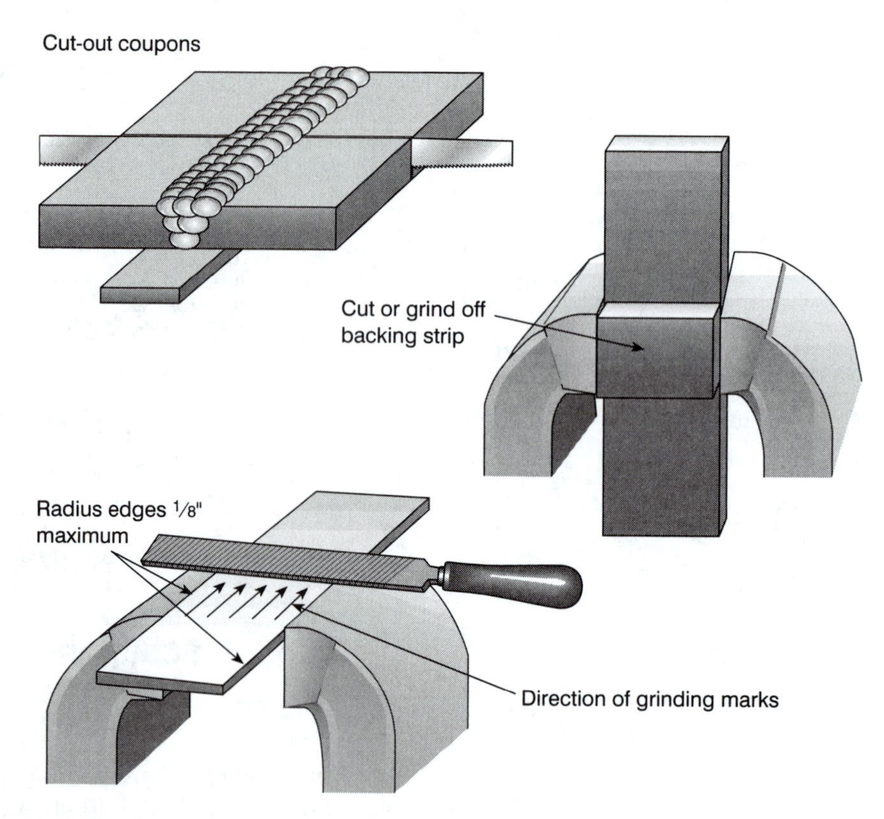

FIGURE B–5 Preparation of coupon for testing.

Testing the Coupons

The welder shall have passed the AWS *Structural Welding Code-Steel D1.1* qualification for guided bend test if the coupon meets the following criteria:

1. No discontinuities of ⅛-inch or more in length, measured in any direction.
2. Sum of all discontinuities shall not equal or exceed ⅜-inch.
3. A crack along the edges that were filed and that contain visible slag shall measure no more than ¼-inch.
4. A crack along the edges of the coupon that were filed and that contain visible slag shall measure no more than ⅛-inch.
5. A coupon with corner cracks greater than ¼-inch shall be replaced with another test coupon from the original weldment.

Passing a guided bend test supervised by a licensed testing faculty will be required for many jobs, and will provide the welder proof of the accomplishment. Many companies require that their welders pass both written and destructive tests before hiring.

Glossary

AC—*See alternating current.*

alloy steel—A steel in which one of the following amounts is exceeded: 1.65% manganese, 0.60% silicon, or 0.60% copper. Or, (2) a steel to which a definite minimum quality of aluminum, boron, chromium, or any other alloying element is added, to achieve a desired alloying effect.

alternating current (AC)—The back and forth motion of electric current, changing directions 120 times (60 cycles) per second.

aluminum—A popular nonferrous metal that has one-third the weight of steel.

American Iron and Steel Institute (AISI)—An organization that classifies types of steel.

American Society for Testing and Materials (ASTM)—A specification forming body that helps to establish quality.

American Welding Society (AWS)—An organization that plays a major role in setting standards used throughout the welding industry.

amperage—(1) A setting on a power source. (2) A measurement in the rate of current flow.

arc—*See welding arc.*

arc blow—An erratic arc that refuses to go where the welder desires and results in sputter and spatter.

arc flash—A painful, but usually temporary, eye condition caused by the light of the welding arc on unprotected eyes.

arc length—The distance of the electrode from the weld pool.

arc strike—The beginning of the arc for welding. An arc strike can lower the weld quality if made on the base metal outside of the welding area.

AWS—*See American Welding Society.*

backing—A material commonly used with groove welds on the back side of the joint.

base metal—The material being welded.

bevel—Removal of an edge to form any angle less than 90° in preparation for welding.

bevel-groove weld—The use of the butt joint with edge preparation given to one of two pieces forming the joint.

boron—A nonmetallic element used as an alloy to increase hardness.

brazing—A group of joining processes in which the filler metal is heated to liquidus above 840° but below the melting temperature of the base metal, and capillary action takes place.

buildup—A filler metal deposited so as to extend above the surface of the base metal.

butt joint—A joint formed with pieces aligned in the same plane.

carbide precipitation—A chemical reaction of carbon with chromium that reduces the corrosion resistance of the steel.

carbon—A nonmetallic element added to steel and cast iron that affects hardness and, consequently, strength.

carbon steel—A general term covering a large range of steels. See low-, medium-, high-, and very high-carbon steel.

cast iron—An alloy of iron that contains carbon (up to 4.5%) plus silicon.

chromium—An element added to steel to increase hardness and corrosion resistance. Chromium is the principal alloying ingredient of stainless steel.

coefficient of linear expansion—A number given to a material as a measurement of its expansion per inch per degree of rise in temperature.

cold lap—Incomplete fusion within a weld where a bridge of melted filler metal has been created without melting the joint root underneath.

complete root penetration—One hundred percent penetration along the entire joint root.

compression strength—The resistance to being crushed.

constant current—The electric output by which the welder can, to a limited degree, control voltage by raising or lowering the electrode.

copper—A metallic element that is readily soldered, brazed, or welded.

corner joint—A joint formed with pieces aligned in different planes, involving the edge of at least one piece.

corrosion resistance—The ability to withstand the stress of chemical reactions without breaking down.

coupon—A dimensioned specimen from a weld or material, used for a destructive test to determine quality.

covering—Fluxing material that coats an electrode.

cover pass—The final layer of a weld, consisting of from one to several weld beads.

crack(ing)—A fracture of the joint that is the result of welding.

crater—An undesirable depression left when a weld bead is stopped or completed but not filled in.

crater crack—A crack originating in a crater.

current—The flow of electricity.

DC—*See direct current.*

DCEN—*See direct current electrode negative.*

DCEP—*See direct current electrode positive.*

deformation—The change a material undergoes due to stress.

destructive test(ing)—A test of weld quality in which the joint is destroyed.

direct current (DC)—Electric current that flows in one direction.

direct current electrode negative (DCEN)—Welding with the electrode attached to the negative terminal and the workpiece connection attached to the positive terminal on the power source.

direct current electrode positive (DCEP)—Welding with the electrode attached to the positive terminal and the workpiece connection attached to the negative terminal on the power source.

discontinuity—A flaw that is not a defect unless it fails to meet the condition established by the welding code.

distortion—A usually uncontrolled change in the shape of metal, resulting from the heat of welding.

downhill—Downward welding with gravity.

ductility—The resistance to deformation after stretching, twisting, or bending.

duty cycle—The recommended percentage of time that a given power source should be under the load of welding.

edge joint—A joint formed when the edges of two or more pieces are brought together parallel to one another.

effective throat—The shortest distance from the weld root to the weld face at the point of a flush contour.

electrode—A consumable or nonconsumable material through which electricity flows to create an arc.

electrode holder—Holds the electrode in position for welding.

electrode lead—A cable that connects an electrode to the power source.

element—Any one of more than 100 substances made up of one kind of atom.

face bend—Stretching the weld face of a coupon to test the weld.

fatigue strength—The resistance to changing forces or loads.

ferrous—Containing iron.

filler metal—Metal added to a joint.

fillet weld—A weld that joins the surfaces of two pieces, usually at right angles to each other.

fillet weld legs—A part of the weld used to determine the size of a fillet weld, measured from the joint root to the toe.

flammable—Easily ignited.

flat bar—Metal more than 3⁄16 inch thick and usually not exceeding 6 inches in width.

flat position—A position in which the plane of welding is from 0° to 15°.

flush contour—A weld face that is a straight line (flat) from toe to toe.

flux—A material that produces a gaseous shield to protect the weld pool and to stabilize the arc, making welding easier. Flux prevents the formation of oxides, helps filler metal to flow, and removes impurities that can affect weld quality.

frequency—The number of electrical oscillations (cycles) per second.

fusion—The melting and mixing together of the pieces that form the base metal with and without filler metal.

grain structure—Configuration of the crystal composition within the metal.

groove angle—The measurement in degrees for the bevel made on the edge of metal in preparation for welding.

groove face—A surface after beveling is completed where pieces of the joint meet.

groove weld—Any one of eight joint designs for the butt joint.

hard-surfacing—Application of specialized filler metal designed to protect the base metal from wear.

hardness—The resistance to penetration or denting.

heat-affected zone (HAZ)—The area along the weld that is affected by the heat of welding.

high-alloy steel—A steel in which chromium, manganese, or nickel content equals 12% or better.

high-carbon steel—A steel that contains 0.45% to 0.65% carbon.

horizontal position—A position in which the plane of welding is from 15° to 100°.

hot-rolled steel—A steel that has been deformed into shape at temperatures that allow course grains to recrystallize.

impact strength—The resistance to sudden force without fracture.

incomplete fusion—Lack of fusion between the weld bead and the base metal.

input—High-voltage, low-amperage electrical power that flows into a power source.

insulator—A device that does not conduct electricity and keeps the electrode holder from becoming part of the arc and short-circuiting against the base metal.

interpass—Time between the laying of the next weld bead.

inverter—A lightweight power source that takes advantage of computer technology; some are versatile enough to be connected to most electric input systems.

iron ore—The raw material used in the manufacturer of cast irons and steels.

joint—The junction of two or more pieces joined together by welding, brazing, or soldering.

joint root—The closest part of the pieces forming a joint.

joint root opening—A separation at the joint root, usually for greater penetration.

lap(ping)—To lap over. The degree one weld bead extends over another weld bead.

lap joint—A joint formed with two pieces extending over each other in parallel planes.

layer—Lapping weld beads on a metal surface.

lead—Cable extending from the power source to the workpiece connection or the electrode holder.

low-alloy steels—A steel in which chromium, manganese, or nickel content is less than 12%.

low-carbon steel—A steel that contains 0.1% (0.001) to 0.3% (0.003) carbon.

manganese—A metallic element added to steel to increase its hardness.

manual(ly)—A term used to describe welding, brazing, soldering, or cutting performed using hand-held equipment.

matensite—A hard, pinlike grain structure that forms when steel cools too rapidly for the pearlite grain structure to form.

medium-carbon steel—A steel that contains 0.3% (0.003) to 0.45% (0.0045) carbon.

melt-thru—Uncontrolled overheating that results in an unexpected hole.

mild steel—*See low-carbon steel.*

mill scale—Oxides (blue-black in color) on the surface of steel that were formed during manufacturing and can affect weld quality.

molybdenum—A metallic element added to steel to increase its hardness and tensile strength.

nickel—An alloying element that increases the hardness of steel.

nondestructive testing—A test of weld quality in which the joint is not destroyed.

nonferrous—Containing no iron.

output—Low-voltage, high-amperage electric power that flows into the welding arc from a power source.

overhead position—A position in which the plane of welding is from 0° to 80°.

oxyacetylene—A welding, cutting, or brazing process that requires oxygen and acetylene.

parameters—Important welding variables.

pearlite—The formation of ferrite (pure iron) and cementite (hard iron carbide) in the grain structure of steel and cast iron.

pipe—A structural, steam- and fluid-carrying medium measured from inside diameter up to and including 12-inch diameter. Pipe over 12 inches is measured by outside diameter.

plate—Metal that is more than 3⁄16 inch thick and usually over 6 inches wide.

plug weld—A weld in a circular hole of one piece joined to a second piece in a lap joint.

porosity—Pinholes in a weld as the result of gas pockets.

postheating—The application of heat to the base metal after soldering, brazing, or welding.

power source—A piece of equipment designed for welding, cutting, or both.

preheating—The application of heat to the base metal before soldering, brazing, or welding.

press brake—Mechanical equipment for bending or shaping metal.

properties—Characteristics of a material such as its hardness and ductility.

quality—Welding that meets a credible standard of excellence.

rectifier—A device that converts alternating current to direct current.

root bend—Stretching the root of a coupon to test the weld.

root face—An edge that has not been removed by a bevel, establishing the boundary of the root opening.

root opening—*See joint root opening.*

root penetration—The depth of penetration by the weld into the joint.

rust—The usually corrosive iron oxides that form on iron and steel.

safety glasses—A critical piece of safety equipment with side shields to protect the eyes from damage relating to welding. Meets ANSI Z87.

scratch method—A technique to initiate welding by dragging the end of an electrode on the base metal, like striking a match.

shear strength—The resistance to force, pushing against the joint from opposite directions.

sheet—Metal that is 3/16 inch or less in thickness and over 6 inches wide.

shielded metal arc welding (SMAW)—An arc welding process that uses an electric arc, usually between a flux-covered consumable electrode and the weld pool.

side bend—Stretching the side of a coupon to test the weld.

silicon—A nonmetallic element that added to steel can increase its hardness and improve its mechanical properties.

slag—Oxidized metal that can affect weld quality.

slag inclusion—Nonmetallic material trapped within the weld.

slot weld—An elongated plug weld.

soldering—A group of processes in which the filler metal is heated to liquidus below 840°F and below the melting temperature of the base metal.

spatter—Molten metal from the electrode thrown away from the weld.

square-groove weld—A weld that uses a butt joint without edge preparation.

stainless steel—A high-alloy steel with at least 12% chromium. Stainless steel is well suited for corrosive applications.

steel—*See low-, medium-, high-, and very high-carbon steel.*

stress—Forces acting on a material.

stringer—A weld bead deposited without weaving.

structural shapes—The forms in which metal can be purchased.

tack weld—A short, temporary weld used to hold a weldment in position to complete welding.

tap method—A technique to initiate welding by striking the end of an electrode on the base metal, raising it quickly.

tee joint—A joint formed with two or three pieces that are usually at right angles to one another.

tempering—A heat treatment by which hardened steel is toughened.

tensile strength—The resistance to being pulled apart.

tensile test—A test used to measure the strength of a material subjected to a force trying to pull the material apart.

toughness—The resistance to fracture from a constant force.

transformer—A power source that uses alternating current as the output for welding.

travel angle—The angle of the rod, electrode, gun, or torch in relation to the direction of travel along the joint.

travel speed—The movement of the weld pool across the joint.

tungsten—An element added to steel to increase hardenability.

ultraviolet light—Invisible light that can damage the skin and is especially harmful to the eyes.

underbead cracking—A defect that can occur in high-carbon and alloy steels in the heat-affected zone.

undercut—Groove left unfilled by deposited metal. Underfill along the toes of the weld.

uphill—Upward welding against gravity.

variable-voltage—A power source with a manual voltage control.

vertical position—A position in which the plane of welding is from 15° to 100° with 10° of pivot.

very high-carbon steel—A steel that contains 0.65% (0.0065) to 1.5% (0.015) carbon.

V-groove weld—A weld in which the edges of both pieces forming the joint are beveled.

volatile—Capable of exploding.

voltage—A measurement of the force causing the flow of electricity.

weaving—Any one of several patterns of movement from side to side during welding.

weld—The melting together of the joint with or without filler material.

weld bead—A weld from a single pass.

welder—A person with the skill and knowledge to make quality welds.

weld face—The surface of the weld, extending from the weld toe on one side to the weld toe on the other side.

weldability—The degree of difficulty by which a metal can be welded successfully.

welding—Any process that joins materials together by heat.

welding arc—A controlled short circuit by which the electric current causes intense heat in a weld pool between an electrode and a base metal.

welding gloves—Gloves designed specifically for welding—not ordinary work gloves. Gauntlets protect the lower arms.

welding position—Can be flat, horizontal, vertical, or overhead.

welding procedure—Detailed instructions that must be followed to achieve quality welding.

weldment—An assembly by welded parts.

weld pool—Liquid metal formed at the point of welding.

weld root—The point at which the weld penetrates into the joint root; the farthest point from the weld face.

weld toe—Runs the length of the weld to establish the boundary of the weld face with the base metal.

work angle—The angle of the rod, electrode, gun, or torch in relation to the base metal.

workpiece connection—Connects the base metal to the power source. Has replaced nonstandard term "ground."

workpiece lead—A cable connecting the workpiece connection to the power source.

Index